THE GENERAL GENETIC CATASTROPHE

THE GENERAL GENETIC CATASTROPHE

ON THE DISCOVERY AND THE DISCOVERER

NILS K. OEIJORD

iUniverse, Inc.
Bloomington

The General Genetic Catastrophe
On the Discovery and the Discoverer

iUniverse books may be ordered through booksellers or by contacting:

iUniverse
1663 Liberty Drive
Bloomington, IN 47403
www.iuniverse.com
1-800-Authors (1-800-288-4677)

ISBN: 978-1-4502-9764-6 (sc)
ISBN: 978-1-4502-9765-3 (ebk)

Printed in the United States of America

iUniverse rev. date: 02/17/2011

In memory of my best friend (and sixth cousin) Anfinn who ended his life 40 years old after 20 years with unimaginable suffering from the gene damage of schizophrenia.

In memory of my first cousin Sigrid who died 40 years old after some 28 years with unimaginable suffering from the gene damage of diabetes type 1.

In memory of my first cousin Astrid who died 25 years old due to gene damage.

Contents

I   Overview

"It is absolute imperative that we protect, preserve, and pass on this genetic heritage for man and every living thing in as good a condition as we received it."
David R. Brower

"All human disease is genetic."
Nobel-laureate Paul Berg

"Genetic theories, I gather, have been cherished academically with detachment."
Allen Tate

"Is migraine a genetic illness? This question was previously controversial, but today the answer yes is generally accepted. The scientific evidence is the significantly increased familial risk of migraine, and the significantly higher concordance rate of migraine in monozygotic than dizygotic twin pairs. Finally, the three identified ion-channel genes that can cause familial hemiplegic migraine provide very strong evidence of genetics."
Neurol Sci. 2008 May; 29 Suppl 1: S52-4

"Genetic diseases in Norway" gave 5 hits on Google, January 1, 2011, but none of the hits addressed genetic diseases."
NKO

A *general genetic catastrophe* (GGC) has been developing during the last some 130 years, but mostly after ca. 1950.

The HIGH and INCREASING prevalence and HIGH and INCREASING incidence of genetic damage/genetic

1

disorders/genetic diseases all over the world logically PROVE that we have a general genetic catastrophe. (The increasing incidence is *explosive*. Examples: Due to new mutations and surviving carriers of damaged genes, it's estimated that the rates of hemophilia, cystic fibrosis and phenylketonuria are increasing every generation by 26 percent, 120 percent, and 300 percent, respectively.)

Science-based technology has polluted the whole world with some 1,000,000 mutagenic chemicals in addition to mutagenic radiation.

Even gasoline vapor and diagnostic use of X-rays permanently damage our chromosomes and genes, and burning cigarettes produce smoke containing 800 mutagens.

Chemical by-products of our scientific civilization, like dioxines, PCBs, and PAHs, are destroying DNA even in the Arctic.

In the USA, officially, already about 4 percent of the babies are now born with a major genetic disease or a major birth defect, and officially more than 1 percent are now born with a major chromosome damage. But what about the 96 percent? Well, they live in the same environment. So, of course, they too are gene-damaged and chromosome-damaged to some extent. But that's not discovered, as yet. And what about the next generation(s)? No doubt, they will all be tragic victims of the exploding GGC. By the way: today, already, more than 10 percent of the population in the Western world has a so called rare (genetic) disease. (Why "rare"?) Totally there are some 15,000 human rare diseases (as yet, only some 7,000 are named) and some 150,000 types of human gene damage. Therefore, the prevalence and incidence of each disease or gene-damage does not need to be particularly high before humanity is genetically destroyed. But science is sleeping. In the Western world today, about 85 percent (see below) of us is tortured and finally killed by a common or rare genetic disease. But the public does not know that.

Tragically, the scientific community is misunderstanding or defining away the GGC *caused by manmade mutagenic pollution.*

I (Nils K. Oeijord) am the discoverer of the general genetic catastrophe caused by manmade mutagenic pollution. I want me to get the recognition I deserve.

I have now earned a place in *Who's Who in the World* (28th Edition), in *Great Minds of the 21st Century* (5th Edition), and in *2000 Outstanding Intellectuals of the 21st Century* (2011 Edition). So I do think I've got that recognition. This little book is written to address the history of the discovery and the discoverer of the General Genetic Catastrophe (GGC) caused by manmade mutagenic pollution

What exactly did I discover?

In the past several famous geneticists and biologists, and others, have written about the damage to natural selection and the resulting degeneration in humans. However, in the past, no one has written about *the worldwide general genetic catastrophe caused by local and global general manmade mutagenic pollution* (mutagenic radiation and mutagenic noise included). Googling *general genetic catastrophe* (even googling *genetic catastrophe* and *derailed evolution*) we find that these expressions are linked to my name and my name only. Also, searching Amazon.com and scientific reports gives the same result. The rest of this book brings more details on the discovery and the discoverer of the general genetic catastrophe.

Ronald A. Fisher and William D. Hamilton (world's two best mathematical geneticists) understood these things very well, but they did not discover that pollutants that are mutagenic and/or carcinogenic are the real problem. Charles Darwin lived in a heavily polluted industrial country but he did not see the threat to our hereditary material. H. J. Muller made mutation an experimental subject by devising an objective way of measuring it and showing that ionizing radiation is mutagenic. Muller spent much energy in a crusade against unnecessary human exposure to radiation. Interestingly, he gave little attention to chemical mutagens. Muller did not discover the general genetic catastrophe. It is almost a miracle that the following big guns did not discover the general genetic catastrophe: F. Galton, W. Weinberg, J. B. S. Haldane, S. Wright, H. T. Odum, J. Watson, F. Crick, R. Franklin, B. McClintock, L. Pauling, R. Carson, Jacques Monod, E. O. Wilson, Jared Diamond, R. Dawkins, S. J. Gould, J. C. Venter.(More names here: http://en.wikipedia. org/wiki/List_of_geneticists.) These extremely talented scientists practiced academic keyhole science not seeing the general and catastrophic world picture. But what about the popular science writers, the science journalists, the statisticians, the professors, the teachers, the health authorities, the health professionals (doctors, physicians, nurses), the politicians, the governments, the church, the pope, the environmentalists, the health prophets, or the doomsday prophets? No, they did not discover the manmade global general genetic catastrophe due to manmade local and global mutagenic pollution.

According to WHO: In Europe in 2010 some 85 percent of all deaths are due to genetic diseases. According to Statistics Norway: The number of cancer deaths in Norway per 100,000 people per year has tripled during the last 50 years (1960 – 2010). I do think these numbers tell us that the general mutation rate has tripled during the last some 50 years. This situation *alone* means that we do have a general genetic catastrophe. Even tobacco smoking (including second hand smoke) *alone* does represent a general genetic catastrophe because *tobacco smoke contains a lot of powerful mutagens*. According to WHO: 443,000 Americans die because of tobacco smoking each year. Second hand tobacco smoke kills about 50,000 of them. More than 50 chemicals in second-hand smoke are known to cause cancer. Smokers are 50 to 100 times more likely to get lung cancer than non-smokers, after exposure to asbestos.

In the five years following 9/11/2001, more than 2,200,000 Americans have died from smoking. Compare that to the 6,000 Americans who died on 9/11 and in the war efforts in the five years following 9/11/2001. More than 20,000,000 Americans will likely die between 2000 and 2050 from the effects of smoking. According to WHO: 1,2 million people in China die because of tobacco smoking each year. That's 2,000 people a day. In the whole world: tobacco smoking will kill 1 billion people in the 21st century. However, *the general genetic catastrophe due to smoking is much more catastrophic than the killing of some 1 billion people in one hundred years time.*

Except for the US population, the trend is that the Western world's population is dying out. Only the USA lies over the replacement level of ca. 2.1 children per mother. The world's total fertility rate is now 2.7 children per mother. In 1950 it was 5.0 children. Fewer children per family causes less poverty. Note that the causal chain goes from fewer children to less poverty, not vice versa. The Western families want two or three children (or more), but they are biologically unable to have them. And in the poor countries of the world they cannot afford birth control pills or condoms. Nevertheless the world's populations are heading for extinction. Why? I believe that the general genetic catastrophe is a part of the answer.

Why was the GGC not discovered earlier?

There are hundreds of answers to this question. Firstly, in the 1800s and the 1900s science was not that advanced as we generally today do believe.

Even today, it seems, most geneticists don't fully know or understand the following facts:

For developing a better or useful organism (or at least an organism in working order) each new mutation (in a coding gene) must be introduced, and tried out (by natural selection) one at a time while all the other genes are kept unmutated. Therefore the average natural (undisturbed) mutation rate must be ca. 1 new coding mutation per new individual, i.e. per fertilized egg. The next generation of the genome must on average have maximum ca. 1 new coding mutation (the same mutation in all cells) relative to the parent's genomes. In order to ascertain whether the mutation is good or bad it is essential that mutations are introduced one at a time, while all the other genes of the individual are kept unmutated.

Mutations are totally random events. Therefore they cannot, in general, be allowed to stay in the general population. Internal and external systems repair and protect our DNA. The internal systems consist of DNA-repair mechanisms, DNA-protection mechanisms, and killing of bad sex cells and fetuses. The external systems consist of the production of a very large number of children, a large population filling the environment, and the killing (by natural selection) of bad genes in newborns, children, and adults. However, *in this book we are not discussing natural selection, breeding, eugenics, and similar subjects.* We are discussing manmade mutagenic pollution and manmade DNA damage, only. *We are discussing the GGC.*

The carcinogenic effects of chemicals were discovered in 1775 in London when chimney sweeps were found to cause cancer of the scrotum. In the 1920s, in Germany, they found that tobacco smoke causes lung cancer.

H. Muller won the Nobel prize in 1933 for the discovery that genes are artificially mutable. (He used radiation as mutagen.) Artificial mutation kick-started genetics in the

1940s by showing that one gene specifies one protein. But already in 1902 A. Garrod conjectured that a gene was a recipe for a single chemical. Actually, Garrod discovered "inborn errors of metabolism."

Marie Curie died of cancer in 1934 due to radioactive radiation. She did not understand that X-rays are biologically damaging. A lot of the early X-ray workers were killed by cancer caused by radiation. Science did not understand that even low-level radiation is damaging. Even today, breast screening with X-rays puts women at risk, but science, in general, doesn't care much. And even today mainstream science does not understand, in general, that low-level radiation is mutagenic. Today, diagnostic use of X-rays is killing some 100.000 humans worldwide per year. (Calculated from the 100 humans officially killed (of cancer) per year in Norway due to X-ray exposure.) However, what is important here is that the survivors too, i.e. the rest of the population, have their DNA damaged due X-ray radiation. (Cancer is caused by gene damage.)

I was six years old in 1953 when Watson and Crick discovered the structure of DNA. (In 2008 I visited the Eagle pub in Cambridge, where Crick stormed in on February 28, 1953, and shouted "We've discovered the secret of life.") Yes, 1953 is the year of the discovery of DNA's structure.

Not until the 1950s was the number of human chromosomes established, and not until ca. 1965 was it possible to look at differences in proteins, the products of genes. Modern genetics began in 1965 when the whole genetic code was known.

In the 1960s B. Ames discovered that chemicals and radiations that caused cancer were very good at damaging DNA.

In the 1960s the leftist view was that schizophrenia (a genetic disease) is a sane response to an insane world.

Asthma runs in families, so it's genetic. But in the 1950s and 1960s many doctors said that *asthma is not a disease.* They even recommended tobacco smoking against asthma. (In the 1990s fifteen "asthma genes" were found.) In the 1870s, Armand Trousseau included a chapter on asthma in his *Clinique Medicale.*

Huntington's disease first became *widely* known in 1967 when it killed Woody Guthrie. But Huntington's disease was first diagnosed by G. Huntington in 1872. He noticed that it runs in families.

Open a modern textbook and you may find that they claim that Alzheimer's disease (first diagnosed in 1906) is *not* a genetic disease!

In the 1970s the scientists, in general, did not accept that cancer is a genetic disease. But in 1979 DNA from tumors was used to prove that damaged genes *alone* could cause cancer.

A 1976 (sic) survey found that a third of British students had never heard of DNA.

It's said that evolution "became genetic" in the 1970s.

Evolutionary Psychology (EP) was born in the late 1980s, and most leftists immediately went to war against the new revolutionary science. This was the leftist Lysenkoism (see more below) one more time, but this time in the field of human behavior.

Gene therapy was born ca. 1990. Biotechnology began in the mid-1990s. In many ways, mostly indirectly, gene therapy and biotechnology are a genetic catastrophe for humans, animals, and plants. Example: Cloning is dysgenic because it avoids sex which mainly evolved to protect the genes and the chromosomes.

The leftists in the Soviet Union, and the genuine leftists in the West hated "bourgeois science." That is they hated, for examples, history, economy, science of politics, sociology, linguistics, semiotics, cybernetics, probability theory, genetics, quantum mechanics (the leftists didn't like the word "random"), theory of relativity (the leftists didn't like the word "relative"). In the 1930s, 1940s, 1950s, and the 1960s, the leftists with their science leader T. Lysenko totally destroyed biology and medicine (including genetics, of course) in large parts of the world, not only in the Soviet Union. So, before ca. 1970, theoretically, no leftist scientist in the East or in the West, was potentially able to discover the GGC.

"...gene is mythical part of living structure which in reactionary theories like Mendelism-Veysmanism-

Morganism determines heredity. Soviet scientists under leadership of Academician Lysenko proved scientifically that genes don't exist in the nature."
(Soviet Encyclopedia, ca. 1950)

"Insisting that this theory corresponded to Marxism, he successfully attracted official support of the Party to his side. He was named president of the Academy of Agricultural Sciences in 1938. He began a persecution of those colleagues who did not agree with his theories, notably the founder of the Academy, Vavilov (who was deprived of work, arrested, and died in the Gulag). [...] [Lysenko] Became dictator in biological sciences under Stalin, whose cult he supported. In effect, he became a Stalinist deputy for science, like Zhdanov for culture, Voroshilov for the army, Beria for everything in the country. [Lysenko] was personally responsible for the exile, torture, and death of many talented scientists, and for an environment of oppression and backwardness in Soviet science."
(Who's Who in Russia and the Former USSR by Terra Moskva, 1994)

"Genetics" and "genes" were tainted words after WWII. And to combine these words explicitly with terms like "altruism" and "aggression" was considered absolute anathema all over the World. Even in the US, in the 1960s and 1970s, there was a climate of suppression, punishment, and defamation of scientists who understood the role of DNA in human behavior and disease. Leftist academic opinion leaders, and even the UNESCO, directly or indirectly, persecuted scientists who fully understood the importance of DNA in human life itself and human society.

However, between about 1950 and about 1990 the political correct (PC) leftist ideology of environmental determinism came gradually tumbling down. Finally, the period of Neo-Lysenkoism was over.

For more than a hundred years, "guilty" parents have endured the environmentalists blaming them for causing schizophrenia, autism, anorexia, homosexuality, etc. The same environmentalists dismissed genetic diseases as "psychological reactions", "false consciousness", etc.

Tragically, however, even today, in the new millennium, leftist environmental determinism is PC in parts of the world, for example Norway. To prove the truth of what I'm claiming just now, I will soon quote an article written by the best Norwegian science writer.

One of the most, if not *the* most, popular Norwegian TV comedian, Harald Eia, discovered in 2009 - 2010 (see the copied/quoted article below) that in Norway leftist die-hard social determinists rule the country, so to speak. They alone (almost) deliver the premises for the politicians and the government. "Genes" and "genetics" are taboo words for these leftist researchers and scientists. When they hear the word "USA" they simply laugh. (See the TV series). It's terrible to watch Harald Eia's TV-series about these untalented and incompetent, but powerful, people in the Norwegian society. They could never ever be able to warn against mutagenic pollution or discover the general genetic catastrophe. The article below is, believe it or not, written by a Norwegian. It's an extremely important historical, social, and scientific document.

The copy (quote) below is from:

http://eusja.wordpress.com/2010/04/26/norway-brainwashed-science-on-tv-creates-storm/

Quote:

"Norway: Brainwashed Science on TV Creates Storm
By EUSJA member.

Bjørn Vassnes, Science Journalist, Norway.

Normally, science is not a subject in Norway. If you ask people on the street, very few can name a single Norwegian scientist, dead or alive. And even the biggest newspapers do not have science reporters, even if Norwegians read more papers than anyone. Then, suddenly, the whole nation is discussing science: in the newspapers, on the TV, in the radio, and most of all in blogs and other internet media. With a temperature that you usually find in much more southernly countries.

No, there has been no big discovery. No controversy over GM food, stem cells or research animals.

SOCIAL SCIENTISTS FELT FOOLED BY TV-COMEDIAN TURNED TO SCIENCE JOURNALIST

The heat is generated by Harald Eia, a TV-comedian turned science reporter, who is exposing social scientists and gender researchers in a not very flattering manner in a TV series called "Brainwashed". The uproar started already last summer, more than half a year before the series was ready. Some social scientists who had been interviewed by Eia, went out in the press to say they felt they had been fooled, tricked to expose themselves by "dubious" tactics.

What Eia had done, was to first interview the Norwegian social scientists on issues like sexual orientation, gender roles, violence, education and race, which are heavily politicized in the Norwegian science community. Then he translated the interviews into English and took them to well-known British and American scientists like Robert Plomin, Steven Pinker, Anne Campbell, Simon Baron-Cohen, Richard Lippa, David Buss, and others, and got their comments. To say that the American and British scientists were surprised by what they heard, is an understatement.

SCIENCES DOMINATED BY IDEOLOGY

In Norway, the social sciences have been more dominated by ideology and fear of biology than in perhaps any other country. This has a long history starting in the 60s. Social science became very much bound up with the ideology of the Social Democrats, who put pride in the fact that Norway was the most egalitarian country in the world. And with the new wealth from the North Sea oil, it became possible to create a society with very little poverty. Which of course has been good for most Norwegians.

MONEY CORRUPTS SCIENCE

But science started to suffer. With so much easy money, few wanted to study the hard sciences. And the social sciences suffered in another way: The ties with the government became too tight, and created a culture where controversial issues, and tough discussions were avoided. Too critical, and you could risk getting no more money.

It was in this culture Harald Eia started his studies, in sociology, early in the nineties. He made it as far as becoming a junior researcher, but then dropped off, and started a career as a comedian instead. He has said that he suddenly, after reading some books which not were on the syllabus, discovered that he had been cheated. What he was taught in his sociology classes was not up-to-date with international research, and more based on ideology than science.

One of the problems, which has prevailed until now, is that the social sciences in Norway not at all will consider biological (evolutionary, genetical) factors in their analyses of human behavior. Even gender roles and sexual identity are explained as 100 percent determined by culture. The theory is that boys and girls are created equal – at least in their heads. All talk about possible inborn differences in interests or capabilities was taboo. Because Norwegians wanted everybody to be equal, it was considered threatening to even ask if there might be some inherited differences. Not only between the sexes, but between people generally.

This was of course a theme in other countries as well at some stage (remember what happened to Larry Summers at Harvard), but in Norway it has been more pronounced than anywhere else (with the possible exception of Sweden). And in Norway this became a big problem because there are few scientists, and most research is sponsored by one source, the Norwegian Research Council, which has strong links with the government.

NO CRITICAL DISCUSSIONS

The situation was such that until recently, there has been no critical discussion of the basic dogmas about sex and gender, about criminality and about the Norwegian school system. Some questions were asked when Norway joined international school tests, and we discovered that we had fallen behind, to a level with much poorer countries. And there was some discussion why the most egalitarian country in the world had bigger differences in choice of education and careers between the sexes, than any other developed country.

This has been called the «gender equality paradox», and nobody could explain it. The common reaction was that we just had to work harder to reach our egalitarian goals. But of course, this "paradox" is easily explained if one takes evolutionary psychology into consideration: Because Norway has such a high living standard that you can live a decent life also with "female" jobs such as nursing, the women now choose careers that suit their psychological needs. But to say such things aloud, was like putting yourself in the gauntlet.

If Eia had presented the series five years ago, he also would have had to try the (media) gauntlet. But even in Norway, the outside world is creeping in, and last year he felt that the time was ripe for this project. He was maybe a bit optimistic, and some of the interviews created such storm, long before the series was aired, that there was a possibility that the project has to be abandoned. Some scientists even threatened to sue him.

But his standing as the most popular TV-comedian in Norway, made it difficult for NRK (the national broadcaster) to back off, and after some delay and bitter dicussions in the media, the series went on air on March 1. It immediately became one of the most watched series on Norwegian TV, and the most watched program on internet-TV.

LOOKS NAIVE, BUT IS WELL PREPARED

For many people, it was difficult to see Eia in his new role as an investigative science reporter (a kind of science journalism's

Michael Moore), but he was well prepared. He could look naive, but he often knew more about the subjects than the scientists he interviewed, which made some of them look like arrogant ignorants. One of them fled the country, declaring that Eia had "ruined her life".

Eias methods have been criticized as being unfair to the Norwegian scientists, but they were given a chance to defend themselves, and his ways of interviewing people are not worse than most politicians or business people are used to. One problem is maybe that the Norwegian scientists had not met any critical journalists before.

But the main problem, which Eia has exposed so brilliantly, is that much of Norwegian social science, and gender science in particular, is built on very shaky ground. Most studies have been done without even considering factors like heredity: The reason why some people turned criminals, or did badly in school, was always explained by social and cultural factors. To even mention heredity as a possible factor, was met with condescending laughter or irritation.

METHODS CRITICIZED, RESULTS JUSTIFIED

Before the series, most of the social science community was very skeptical, but now even established scientists have admitted that the critical light had been justified. Another effect of the series has been that scientists you almost never heard from in the public: psychologists, biologists and other natural scientists, have started to write in newspapers and participating in debates.

So even if Eia's methods have been criticized, there is now a general agreement that the result of this project has been good for both the sciences and society as a whole. For the first time, science is really being discussed. Even if many strange things have been said and written, this has been (and still is) a unique educational process for both the general public and the scientific community.

The series "Hjernevask" (Brainwashed) can be watched, in Norwegian, here: http://www1.nrk.no/nett-tv/klipp/625996

EUSJA tried to help Norwegian science journalists and science communicators in the past to create an association of science journalists. In vain. Norway is, apart form small Iceland, the only one of the technically and scientifically highly developed Scandivavian countries where science journalists are a very rare species. What happens to a country, were science and technology are not publicly discussed, shows dramatic and amusing story by Bjørn Vassnes, one of those rare species in Norway. It shows once again that science journalism is important for a democracy, and that science communication, science PR on the contrary can ruin the credibility of science. Norway needs a science journalists' association. / Hanns-J. Neubert"

Unquote.

To fully answer the question "Why was the GGC not discovered earlier?" we have to look at much older history as well.

The 1600s and the 1700s were the times of Descartes, Newton, Leibnitz, Kant, ... It was the times for discoveries, inventions, and studies. They studied everything: from molecules, via the human body and brain, to planets (mostly planet Earth) and stars. They developed medical science, universities, hospitals (including hospitals for mentally ill persons). In the 1700s Carl von Linne gave names to the species of the Earth. Of course, *they gave name to almost all human diseases and most human anomalies that existed at that time, and almost everything was carefully written down.*

But nearly all of today's some 7,000 described genetic diseases were first diagnosed and described in the 1800s or 1900s. Conclusion: *Science must stop explaining away today's genetic diseases by claiming that these diseases have always existed.*

Genetic diseases are generally lacking in local history books, family history books, old classic history books, old literature, the holy books, old medical textbooks, and old medical journals. (See much more on this below.)

The old names described the owners physically or mentally. Examples: Ludovicus the skinless, Erik the Red, Eirik Bloodaxe, Inge the Hunchback, Harald Finehair, Haakon the Good, Magnus the Good, Magnus the Blind, Haakon the Young, Cnut the Great. There are, I think, relatively few old names pointing to genetic illness. The names Ludovicus the Skinless and Inge the Hunchback point to genetic disease. But the names Magnus the Good and Cnut the Great, for example, do *not* point to genetic disease.

On names of old time illnesses, see:
http://www.homeoint.org/cazalet/oldnames.htm
http://www.michiganancestry.com/files/illnesschart.pdf.
Note that there are relatively few names of genetic diseases on the lists.

We, who live today, have a good overview of the last 4 – 5
generations, and we can watch directly that we do have an
explosion of genetic diseases during these 4 – 5 generations. But
this very fact is not science, you know, so therefore science simply
does neglect the whole thing. "We need more research."

In the 1800s Europe and North America went *from witch-*
hunt to disease-hunt.

"Nor were witches secret pagans serving an ancient Triple
Goddess and Horned God, as the neopagans claim. In fact, no
witch was ever executed for worshiping a pagan deity. Matilda
Gage's estimate [1893] of nine million women burned is more
than 200 times the best current estimate of 30,000 to 50,000
killed during the 400 years from 1400 to 1800 — a large number
but no Holocaust. And it wasn't all a burning time. Witches
were hanged, strangled, and beheaded as well. Witch-hunting
was not woman-hunting: At least 20 percent of all suspected
witches were male. Midwives were not especially targeted; nor
were witches liquidated as obstacles to professionalized medicine
and mechanistic science."
Sandra Miesel

Historically, animals and humans do mob mentally and/or
physically anomalous individuals to death. If authorities were
too slow, or simply did not exist, ordinary people lynched
"suspected" individuals, even neighbors. Physically and/or
mentally anomalous newborns were put in the woods to
die. Darwin named the totality of all these actions: Natural
selection. Simply the dirty work of evolution. Lions, bears,
warriors, etc. were other actors of this dirty play. Witch-
hunting in Europe increased dramatically in the 1600s

during the rationalist age of Descartes and Newton. Why? Why not?

Mutagenic pollution had exploded. Whole cites burned down in Europe. Mutagenic smoke everywhere. London burned down in 1666. More genetic damage. More physical and mental illness. More strange individuals to fear. Simply more witches to take care of. However, only 30,000 - 50,000 anomalous human individuals during 400 years in the whole of Europe is *extremely* few. (Today about 25 percent of the population in Europe and the US do have mental problems, and about 85 percent of the population is killed by genetic diseases.) Then came the 1700s and 1800s, and Christianity and other forces gradually stopped the whole witch thing. I do think these numbers tell us that the genetic damage of mental disease and the genetic damage of physical disease were *rare* before ca. year 1800. An enormous competition has taken place the last some 200 years among doctors and others to be the first to find a *new* (and, back then, extremely rare) disease. *Europe and North America went from witch-hunt to disease-hunt.* Scientists (often priests, back then), doctors, and even ordinary people gradually understood that the strange individuals simply were sick, not evil, not dangerous, not witches. Biologically speaking, today, about 85 percent (see above) of the population are "witches" and "condemned" to death. But first there is torture (example: pain from cancer). However, long time torture (example, again: pain from cancer) has replaced short time torture (example: death from burning).

Each so-called *Holy man* in China is able to recite his pedigree for more than a thousand years into the past. The

Holy men have not, as far as I know, reported about a genetic catastrophe before the 1800s.

In the Mormon Archive inside Granite Mountain (USA) are stored several hundred million names of dead family members, most born after 1500. This is the most extensive collection of family records in the world. The Mormons have not, as far as I know, reported about a genetic catastrophe in the 1500s, 1600s, 1700s, or the 1800s.

But there are thousands of other and better places where old and new information about human individuals are found. Here follows an incomplete list: legends, diaries, family history books, private archives, local history books, census papers archives, historic counties' tax records archives, midwife archives, birth archives, genealogical archives, school archives, military archives, church books, church archives, cemetery archives, state archives, law archives, prison archives, death penalty archives, euthanasia archives, eugenics archives, inquisition archives (open), Stalin archives (open), Nazi archives, Hitler archives, Holocaust archives, medical journals archives, archives of people's mental health, journals of medicine.

In the 1800s families began to take family photos. The families of our grand grand grand parents are frozen in time on these photos. Study their faces and bodies, and try to find a lot of genetic illnesses. I'm confident that the result will be negative.

We can study human remains from the past and the present hunting for damaged DNA. I think I know what we will find.

Humans have always been clever discoverers. Arthritis was discovered ca. 4500 B.C. (Today more than 100 types of arthritis are known.) Migraine was discovered ca. 3000 B.C. Epilepsy was discovered ca. 2000 B.C. (Today epilepsy is causing mass death. Today 1 out of 22 newborns will get epilepsy.) Angina pectoris was discovered ca. 550 B.C. (Today angina is causing mass death.) Celiac disease was discovered ca.150 B.C.

They were clever at discovering new genetic diseases in the 1500s. Example: Prostate cancer was discovered in 1536. They were clever at discovering new genetic diseases in the 1600s. Example: Patau syndrome was discovered in1657. They were clever at discovering new genetic diseases in the 1700s. Examples: Scrotal cancer and schizophrenia were discovered in 1775 and 1797, respectively. They were clever at discovering new genetic diseases in the early 1800s. Examples: Goldenhar disease, hemophilia, and MS were discovered in 1800, 1828, and 1838, respectively.

On the other hand, there are thousands of examples that show that we were relatively bad at discovering new genetic diseases in the 1900s. And today, an enormous amount of gene damage is undiagnosed. More gene damage is undiagnosed today than ever before.

According to *Science*, 8 January 1971, "at least 1,500 diseases are considered to be of genetic origin." Today, 40 years later, we believe that the *total* number of genetic diseases is some 15,000, and that there *totally* are some 150,000 types of gene damage at work in the human population. In reality, therefore, we do have some 150,000 genetic diseases, not 1,500. A hundred times more. Medical science has totally lost control.

The word "children" is extremely prevalent in literature on genetic disease. ("Gene site found for children's food allergy." "Food allergy in children.") Why? Well, obviously, because the prevalence and incidence of genetic diseases are much higher in children than in the total population. Moreover, new diseases are continuously discovered in newborns. This is simply the ugly face of the GGC.

The accelerated decrease in inbreeding (and increase in outbreeding) in Europe and the rest of the world in the 1800s and 1900s has camouflaged the GGC. There are many times more damaged genes out there today than the prevalence and incidence of genetic diseases are displaying for us. These hidden damaged genes are planning to destroy future generations.

A damaged gene goes to the next generations. What happens then? Let's take an example from a pedigree in Virginia, USA. In the start-generation there was only one damaged gene, in one individual, causing genetic deafness. In the next generation there were three genetically deaf individuals. In the next six generations the numbers of deaf people were, respectively, 5, 6, 6, 8, 9, 13. This example proves that one, and only one, damaged gene in one, and only one, individual, causes a genetic catastrophe in the future. Moreover, *in each individual*, since a damaged gene, once introduced, will remain in the genome, we get an increasing burden of gene damage with each succeeding generation because each generation adds new gene damage to those handed down from the past.

When I studied the GGC in Norway I learned that the Norwegian medical geneticists and Statistics Norway were extremely uninformed about genetic diseases. (The situation is not much better in the rest of the world.)

After I wrote the seven letters to Norwegian health authorities in 2002 (see section VI 9) things started to happen. For example: New Norwegian internet sites were published (and new sections of already existing sites were also published) giving information and statistics about genetic diseases in Norway. But these sites are bad, incomplete, unsystematic, or under construction. See, for example: www.rarelink.no and www.fhi.no. For a better (but not good) site, see, for example, www.wrongdiagnosis.com/g/genetics/prevalence. htm, where I found parts of the information given below.

Norway's population is only 4,900,000
In Norway the number of births per year is some 60,000

According to WHO: In Europe in 2010 some 85 percent of all deaths are due to genetic diseases.

According to Statistics Norway: The number of cancer deaths in Norway per 100,000 people per year has tripled during the last 50 years (1960 – 2010).

Diabetes explosion in Norway: New research from the Norwegian Diabetes Association measures explosive growth (some 400 percent) over the past 50 years. Norwegian Diabetes Association is worried about the explosive growth and expects a further doubling over the next 20 years if nothing is done.

The prevalence of dementia in Norway today is ca. 70,000. It's believed that this number in 2030 has increased to ca. 120,000. Alzheimer's disease is the most common form of dementia. Note that the prevalence of Alzheimer's disease in the oldest people is probably about 50 percent.

Every year some 500 children with congenital heart defects are born (live births) in Norway.

I do think the numbers above tell us that the general unnatural mutation rate has tripled during the last 50 years.

Since ca. 1920 the prevalence and incidence of all genetic diseases in Norway and the rest of the world are increasing steadily or explosively.

First, a few more words about the most common diseases in Norway:

Bechterew: 90,000 living with bechterew
Cancer: 200,000 living with cancer
Cancer: 26,000 new cases per year
Cancer: 11,000 cancer deaths per year
Cardiovascular disease: 410,000 living with cardiovascular disease
Cardiovascular disease: 14,000 deaths per year

Chronic pain: 810,000 (World record)
Chron's disease: 2,900 is living with Chron's disease
Chron's disease: 900 new cases per year (WR)
Diabetes 1: 25,000 living with diabetes 1 (WR)
Diabetes 1: 600 new cases per year (WR) (exploding)
Diabetes 2: 175,000 living with diabetes 2
Diabetes 2: 6500 new cases per year
Diabetes (1 and 2): 1200 diabetes deaths per year
Epilepsy: 40,000 living with epilepsy
Epilepsy: 2000 new cases per year
Rheumatic disorder: 300,000 living with rheumatic disorders
Schizophrenia: 10,000 living with schizophrenia
Stroke: 12,000 new cases per year
Stroke: 5,500 stroke deaths per year

And now, let us study (see the list below) only a tiny fraction of the "uncommon" some 7,000 (or some 15,000) genetic diseases in the Norwegian population. Note the many "(sic)."

Year of first discovery or first diagnosis is given in parentheses. The incidence is given as a fraction with 1 as the denominator. The incidence is often calculated by extrapolation from numbers for the USA or other countries. The definition of incidence here (see below) is perhaps a bit unusual, but it gives a much better understanding of the dynamics of the GGC than other definitions do. Incidence = (number of new diagnosed cases in newborns per year + number of new diagnosed cases in non-newborns per year) / (number of newborns per year)

Aagenaes syndrome	(1968)	1 :	100,000
Aarskog-Scott syndrome	(1970)	1 :	50,000
ADHD	(1845)	1 :	20? (sic)
Alagille syndrome	(1969)	1 :	70,000
Albinism	(1908)	1 :	20,000
Allergic rhinitis	(?)	1 :	10 (sic)
Allergies	(?)	1 :	5 (sic)
Alport syndrome	(1927)	1 :	50,000
Alzheimer's disease	(1906)	1 :	55? (sic)
Amytrophic lateral sclerosis (ALS)	(1939)	1 :	800
Angelman's syndrome	(1965)	1 :	15,000
Anorexia nervosa	(1870s)	1 :	50? (sic)
Anxiety disorders	(1900s?)	1 :	10? (sic)
Apert syndrome	(ca.1900)	1 :	120,000
Arthritis (today over 100 types)	(4500 B.C.)	1 :	100 (sic)
Asthma	(?)	1 :	20? (sic)
Atherosclerosis (rediscovered 1950)	(1850)	1 :	55? (sic)
Atopic eczema	(?)	1 :	15 (sic)
Atresia	(1900s)	1 :	5,000
Angina pectoris	(ca. 550 B.C.)	1:	65? (sic)
Aniredia	(1900s)	1 :	30,000
Autism	(1938)	1 :	110 (sic)
Bechterew's disease	(ca.1890)	1 :	400 (sic)
Beckwith-Wiedermann syndrome	(1964)	1 :	15,000
Bipolar disorders	(?)	1 :	40? (sic)
Bladder cancer	(1800s)	1 :	55 (sic)
Bloch-Sulzbergers syndrome	(1900s)	1 :	15,000
Bulemia	(1979)	1 :	50? (sic)

Celiac disease	(1856)	1 :	200
CFS	(ca. 1985)	1 :	100 (sic)
Choanal atresia	(ca 1800)	1 :	12,500
Chron's disease	(1904)	1 :	65? (sic)
Chrouzon disease	(1912)	1 :	100,000
Congenital heart defects (in live births)	(?)	1 :	95 (sic)
Cystic fibrosis (Tay-Sachs disease)	(1930s)	1 :	3,000
Deliberate self-harm (DSH)	(1960)	1 :	20? (sic)
Dementia	(1901)	1 :	50? (sic)
Down syndrome	(1866)	1 :	800
Duchenne muscular dystrophy	(1860s)	1 :	3,000
Dyscalculia	(1919)	1 :	25 (sic)
Dyslexia, severe	(1887)	1 :	10 (sic)
Dystonia	(1900?)	1 :	2,000
Ektodermale dysplacia	(1900s)	1 :	100,000
Epidermolysis bullosa	(1900s)	1 :	100,000
Epilepsy	(2000 B.C.)	1 :	22 (sic)
Exstrophy of the bladder	(1900s)	1 :	35,000
External genital abnormalities	(?)	1 :	4? (sic)
Fabry's disease	(1898)	1 :	30,000
Factor V defiency	(1947)	1 :	?
Factor VII deficiency	(ca.1950)	1 :	70,000
Familial polyposis, autosom. rec.)	(1951)	1 :	4,000
Fibromyalgia	(1975?)	1 :	50
Food allergies	(?)	1 :	20 (sic)
Fragile-X syndrome	(1970s)	1 :	1,500 (m)
Freeman-Sheldon syndrome	(1938)	1 :	400,000
Galactosemia	(1900s)	1 :	100,000

Gaucher's syndrome	(1882)	1 :	50,000
Genital anomalies	(?)	1 :	5,000
Goldenhar disease	(1900)	1 :	5,000
Hemophilia	(1828)	1:	40,000
Hereditary hemorrhagic telangiectasia	(ca.1900)	1 :	5,000
Hereditary nonpolyposis colon cancer	(1900s)	1 :	250
Huntington's disease	(1842)	1 :	4,000
Ichthyosis	(1900s)	1 :	180,000
Immunedefiency	(1800s)	1 :	7,000
Kidney cancer	(1883)	1 :	60,000
Klinefelter's syndrome (XXY)	(1942)	1 :	500 (sic)
Laurence-Moon-Bardet-Biedel syndrome	(1866)	1 :	120,000
LCAT defiency	(1967)	1 :	?
Leukemia	(1845)	1 :	20,000?
Liver cancer	(?)	1 :	5,000
Major depression	(?)	1 :	12? (sic)
Maple syrup urine disease	(1954)	1 :	180,000
Marfan's syndrome	(1896)	1 :	10,000
Mastosytosis	(?)	1 :	100,000
Malignant melanoma	(?)	1:	10 (sic)
Methylmalonic acidemia	(?)	1 :	50,000
Migraine (genes newly found)	(3000 B.C)	1 :	9 (sic)
Multiple sclerosis (MS)	(1838)	1 :	700 (sic)
Muenke syndrome	(1900s)	1 :	34,000
Netherton syndrome	(1958)	1 :	200,000
Neurofibromatosis	(1882)	1 :	3,000
Niemann-Pick disease	(1914)	1 :	120,000
OCD	(1963)	1 :	40? (sic)

Oesofaguesatri	(1900s)	1 :	5,000
Osteogenesis imperfecta	(1800s)	1 :	10,000
Osteoporosis (f)	(1770s)	1 :	3 (sic)
Osteoporosis (m)	(1770s)	1 :	10 (sic)
Parkinsons disease	(1817)	1 :	90? (sic)
Patau syndrome	(1657)	1 :	5,000
Pfeiffer syndrome	(1900s)	1 :	100,000
Phenylketonuria	(1934)	1 :	10,000
PMDD (PMS, severe variant)	(1700s?)	1 :	20? (f)
Prader-Willis syndrome	(1956)	1 :	10,000
Propionic academia	(1900s)	1 :	100,000
Prostate cancer (males)	(1536)	1 :	15 (sic)
Psoriasis	(1841)	1 :	25 (sic)
Retinoblastoma	(1864)	1 :	250 (sic)
Rett's syndrome	(1954)	1 :	10,000
Saethre-Chotzen syndrome	(1931)	1 :	85,000
Schizophrenia	(1797)	1 :	100
Scrotal cancer	(1775)	1 :	?
Sickle-cell anemia	(1870s?)	1 :	500
SLI	(1900s)	1 :	20? (sic)
Sotos syndrome	(1964)	1 :	5,000
Thyroid cancer	(1880s?)	1 :	200 (sic)
Tourette's syndrome	(1885)	1 :	500
Treacher Collins syndrome	(1900)	1 :	10,000
Tiple-X syndrome (XXX)	(1959)	1 :	1,000 (f)
Turner's syndrome	(1938)	1 :	2,500 (f)
William's syndrome	(1961)	1 :	8,000
Wolf-Hirschhorn syndrome	(1961)	1 :	50,000
XYY syndrome	(1956)	1 :	1,000 (m)

My research in 2010 shows that the average life in Northern Norway is ca. 4 years shorter than officially believed. The research shows that the after-WW II-generation is rapidly going away, and that people born in the 1910s and 1920s (before the mutagenic pollution explosion) do still contribute significantly to better the average. But when these people are gone, the average life will become much shorter. People born in the 1930s and later are born in a highly mutagenic and carcinogenic environment. Here is a typical example from 2011: The six funeral announcements in the local newspaper Nordlys, January 14, 2011, have the following birth years: 1917, 1919, 1920, 1920, 1953, 1955. The average here is ca. 1931, while the officially claimed average is ca. 1929. Obviously, the average is covering up the GGC. But if we look at the 1953 and 1955 cases, then we can clearly see the footprints of the GGC.

More disease incidence statistics, see:

http://health-articles-2010.co.cc/heart-disease/alarming-statistics-on-heart-disease-what-you-can-do-about-it.html

www.txalzresearch.org/index.php?option=com_content&view=article&id=52&Itemid=68

www.who.int/genomics/public/geneticdiseases/en/index2.html

www.emro.who.int/ncd/

www.who.int/genomics/professionals/international_organisations/en/

www.paho.org/common/Display.asp?Lang=E&RecID=7092

www.who.int/genomics/anomalies/en/Chapter06.pdf

www.who.int/genomics/about/en/1-4.pdf
www.paho.org/common/Display.asp?Lang=E&RecID=7248

http://en.wikipedia.org/wiki/Rare_disease

www.wrongdiagnosis.com/

www.orpha.net/consor/cgi-bin/index.php

www.ncbi.nlm.nih.gov/omim

www.health.state.ny.us/diseases/chronic/basicstat.htm

www.tdrdata.com/

www.youtube.com/watch?v=m6SzGOtlhzo

http://ideas.repec.org/p/iza/izadps/dp4992.html

http://mpkb.org/home/pathogenesis/epidemiology

www.lifescienceintelligence.com/market-reports-page.
php?id=A701

http://jhsrp.rsmjournals.com/cgi/content/abstract/13/1/53

http://researchandmarkets.net/reportinfo.asp?report_id=16357

Disease increase quotes, see: www.whale.to/m/quotes29.html

However, in reality we do not need to know any genetic disease statistics at all, because the manmade mutagenic pollution explosion must necessarily produce a worldwide general genetic catastrophe (GGC). But my research since 1999, which is outlined in this book, does show *directly* that we do have a global general genetic catastrophe, due to manmade general mutagenic pollution. *The GGC began in the 1700s, increased in the 1800s, and exploded in the 1900s.* (See, for example, the Norwegian situation above.)

It's well known that health care costs have increased dramatically over the past century due to greater prevalence and incidence of genetic diseases. The last boost of the GGC started ca. 1960. Now we have increasing problems with co-diseases, and multiple genetic diseases in the same victim. Examples: "diabetes with depression", "disease X combined with co-diseases such as X and Y".

Using the available genetic disease statistics of today, can you calculate how many humans will be gene-damaged during the next few generations? Do you think our scientists and researchers have already made this calculation? If your answer is "yes", where is it published? If your answer is "no", what do *you* think is the reason why science being that uninterested?

The best psychological explanation why science is neglecting the general genetic catastrophe is found in William James's instinct theory. The *suppressing-thoughts instinct* brilliantly explains why scientists avoid the GGC. This particular instinct is a mighty force. Here is an interesting example:

The Norwegian (national) health director Torbjorn Mork (1928 – 1992) had a PhD on lung cancer and smoking, but he was a heavy smoker and died of lung cancer due to tobacco smoking. Addiction? Yes. The *suppressing-thoughts instinct*? Yes.

By the way, it's a well known fact that, throughout his life, Albert Einstein (1879 – 1955) was one of the greatest supporters of tobacco smoking in the world, and that one of his sons became a schizophrenic.

II An Essay on the GGC (3. ed)

Nils K. Oeijord

"We [the human race]|do not have much time to prove that we are not the product of a lethal mutation."
Science 263: 181, 1994

"I almost think it is the ultimate destiny of science to exterminate the human race."
Thomas Love Peacock

The general genetic catastrophe consists of four major genetic catastrophes - those of cancer, vascular disease, musculoskeletal disease, and behavioral disease. There are identified and described approximately 300 common genetic diseases, and approximately 7,000 "rare" genetic diseases.

The list of gene/genetic damage is growing daily. (The total number of types of gene damage is some 150,000.) The *natural* rate at which mutations and genetic damage occur is reasonably constant. The larger the gene the more common genetic diseases of the gene are. Hence, common genetic diseases are caused by mutations/damage to very large genes.

Gene/genetic damage is far more common than generally understood. There is insufficient recognition of the magnitude of the general genetic catastrophe. The concept of genetic disease has expanded during the last twenty years. The concept has now expanded to be virtually all encompassing.

Even infectious diseases have some relationships to our genes. Nobel-laureate Paul Berg said "all human disease is genetic."

The *natural* mutation frequency is about 1 new mutation in coding genes per generation (per fertilized egg). (The same mutation in all cells.) *This value proves that present-day mutagenic manmade pollution is an absolute genetic tragedy because this value lies near the critical value that separates evolution from extinction.*

The strong increase in the gene damage of genetic diseases like cancer, vascular diseases, Alzheimer's, Parkinson's, diabetes, osteoporosis, obesity, etc, is perhaps the best illustration of the magnitude of the general genetic catastrophe. Certainly the victims of genetic damage are the victims of modern human activities (including science and science based technology).

Some 85 percent of us are tortured and finally killed by cancer, vascular diseases, other non-rare genetic diseases, and "rare" diseases. Science, in general, does not speak about a general genetic catastrophe. On the contrary, by using words like syndrome, disorder, disease, illness, defect, deficiency, failure, etc, instead of gene/genetic damage, science, in general, is defining away and covering up the general genetic catastrophe.

In the US, a rare disease is defined as "one that afflicts no more than 200,000 people." However, more than one out of 10 Americans suffers from just such a rare disease. And some 7,000 "rare" diseases have been described, according to the US National Organization for Rare Disorders. If (when) these 7,000 diseases reach, on average, the 200,000 people level, then each American, on average, will suffer from more than four rare diseases. Most rare diseases are genetic diseases caused by gene damage.

I our globally polluted world our genes suffer damage just through day-to-day living. Our DNA is constantly under attack from mutagenic chemicals and radiation causing a steady build-up of heritable gene damage, in spite of the best efforts of our DNA repair enzymes and other repair mechanisms, such as the SOS repair mechanism, the cell-suicide mechanism, sterility, infertility, natural and unnatural abortion, recurrent miscarriages, death at an early age, and natural and unnatural selection.

There is no evidence that there is a dose below which there is not a mutagenic effect. A small dose of chemicals/radiation to a large population does more harm to the gene pool than a large dose to a small population. In fact, the effluents causing the gene damage currently meet all environmental standards, and often the gene damage-causing chemicals are present at non-detectible levels. We don't know what's actually happening at the genetic level. Besides, genetic damage is particularly insidious because it can take several generations for the effects to show up. Moreover, the total exposure to mutagenic chemicals and radiation is unknown.

The blood of Inuit people and polar bears contains a large number of mutagenic chemicals in high concentrations.

Certainly, the *romantic* notions of "wilderness", "health", "science", and "technology" are outdated.

Although the individual is often powerless to avoid exposure to widely used chemicals, there are examples of mutagens to which people voluntarily expose themselves, for example cigarette smoke.

Unfortunately, mutagenic chemicals occur in widely divergent chemical groups, ranging from simple compounds such as formaldehyde to complex ones such as alkaloids. With every breath of cigarette smoke, the body is confronted by more than 800 mutagenic chemicals which include dioxin-like compounds.

Apropos of dioxins, the gene damage of dioxins is absolutely devilish: The DNA double-helix acts like a zipper, which can open and close. A dioxin molecule acts like a thread-like object which is preventing the zipper from being closed! All mother's milk (human and non-human) of the world contains dioxins, now and in the future! (PCB etc, act similarly.)

Some 200,000 (an increasing number) American infants are born annually with serious birth defects, which include brain anomalies and cleft palate. These some 200,000 infants are probably born in our average environment. "Some 50 million Americans can't read" because of the gene damage of dyslexia. An individual may be super-intelligent but he/she is unable to learn to read because of specific gene damage. These 50 million (an increasing number) individuals are probably born in, and live in, our average environment. The genetic damage of cancer kills approximately 30 percent of the total US population, while the genetic damage of cardiovascular diseases kills some 30 percent of the US population. A 20-year-old individual may suddenly die of a cardiovascular disease. Baby heart disease, baby cancer, and baby Alzheimer's are now well known genetic diseases.

Because every human individual has about 35,000 functional genes, there is an endless number of possible heritable genetic diseases. Furthermore, there is an endless number of possible chromosome anomalies such as chromosome breaks. Ionizing radiation is extremely efficient at causing chromosome breaks. Even gasoline vapor and high-voltage fields cause chromosome breaks. These messy breaks are difficult or impossible for cells to repair correctly. Evidence shows that here the cell's repair-system is fallible even when it is confronted only by a minimal challenge.

Even relatively low (diagnostic) doses of X-rays cause gene damage. Data strongly suggest that preconceptional exposure of the mother to diagnostic doses of X-rays increases the risk of offspring with Down's syndrome. New research results show that diagnostic doses of X-rays are both mutagenic and carcinogenic.

Chemicals in the general pollution damage crucial DNA repair genes. Besides, heavy metals from background pollution damage repair enzymes (change their form and function). When a repair-gene is damaged, the gene damage will magnify the consequences of the cell's subsequent exposures to all mutagens (chemicals and radiation), because of the cell's diminished ability to repair gene damage correctly. And free radicals and other mutagens attack our DNA all the time.

Cells somehow sense gene damage and activate a suicide program, called apoptosis, to kill themselves so that gene damage is not perpetuated. This is the ultimate "brake" against cancer as well. The p53-gene is known to trigger apoptosis in response to gene damage in a cell. If the p53-gene is damaged, then an important way of preventing gene damage from being passed on to the next generation is lost. How can we stop the gene/genetic damage explosion?

Genetic diseases appear to be identical across species. The elimination of genetic diseases in animals can only be accomplished through selective breeding *in a clean environment*. This is not a theory, this is a fact. However, as we have shown above, the situation for our genes is enormously difficult. It looks as if the genetic diseases will be the winner of the race.

Today twice as many dies of diabetes (per 100,000 people) as before insulin was introduced. *Today, in general, twice as many dies of a genetic disease (per 100,000 people) as before a medicine was introduced.*

The genetic damage to human sperm cells is so enormous that we can recognize directly, by looking through a light microscope, that more than 50 percent of the cells are abnormal. There is perhaps nothing that can prevent all sperm cells to become abnormal. After all, normal cells live in the same polluted environment as the abnormal cells.

Of course, the genetic damage to egg cells is even more dramatic because they are relatively few in number.

Men born with a birth defect have a doubled risk, compared with other fathers, of having a child with a birth defect, a large population study revealed. Note that this situation, according to simple mathematics, leads to a gene damage explosion. The genetic catastrophe of breast cancer develops extremely fast: Of all the women with breast cancer, only a tenth have family histories of the disease, and half of this group has a heritable gene damage causing breast cancer!

Department of Education estmates that 20 percent of Americans are learning-disabled. This result is obviously wrong, because the dyslexics (20 percent of Americans) are not the only learning-disabled people in the US. The number of some 7,000 for rare diseases in the US is also wrong. Behind each of the some 7,000 diseases there are lots of different types of gene damage.

It has been estimated that more than 50,000 chemicals are in common use in the United States. Most of these chemicals have not been tested for mutagenicity. Of those that have been tested for mutagenicity, about 20 percent produced mutations in the Ames test. Of the several million chemicals in "uncommon" use in the world, perhaps as much as about 1,000,000 are causing gene damage (mutations). These chemicals are working in your cells at this very moment.

The natural (evolutionary) mutation rate is about one mutation per fertilized egg. If the mutation rate seriously exceeds one mutation per fertilized egg (it now obviously does), then evolution is derailed. Also, if the mutation rate is one, or slightly less than one, mutation per fertilized egg, and, if, at the same time, natural selection is disturbed (now it is), then evolution is derailed? And if evolution is derailed, then a general genetic catastrophe is unavoidable.

Behavior is a biochemical event. Behavioral disorders often have a genetic basis. Most neurobehavioral syndromes are gene damage. With regard to survival of the human species, damage to genes for behavioral traits ("behavioral genes") is much more dangerous than damage to genes for physical traits. A large and rapidly increasing percentage of the population suffers from genetic damage of behavioral disorders such as learning disabilities, depression, antisocial personality disorder, Tourette's syndrome, schizophrenia, bipolar disorder, eating disorders, attention deficit disorder, hyperactivity, fragile-X syndrome, autism, Lesch-Nyhan syndrome, Down's syndrome, etc. If (when) only 100 genetic behavioral disorders reach, on average, the 1 percent level, then humanity is destroyed?

III Books on the General Genetic Catastrophe

Natural selection has used mutations for building up well-integrated organisms. New mutations are likely to upset this balance and are therefore mostly harmful or lethal. ENCYCLOPEDIA BRITANNICA

Human Instincts Explained (2000)

This in-depth focus on genes and instincts, which the author finds surprisingly absent from science and philosophy, enables us to see patterns of the future. The nature of intelligence and the structure of the mind are each contemplated, as well as the nature of war and peace, and the origins of ethics. *Moreover, this book constitutes the first serious message of the general genetic catastrophe and the possibility of derailed evolution.*

Genetic Catastrophe! Sneaking Doomsday? (2002)

This book constitutes the second (see above) serious message of the general genetic catastrophe and the possibility of derailed evolution. Moreover, this book contains the first dictionary of genetic damage to be published.

A Dictionary of Genetic Damage (2003)

This dictionary is the first dictionary of genetic damage to be published. *A Dictionary of Genetic Damage* is provided for spelling reference, categorization reference, and information on the enormity of the genetic catastrophe. Note that *A Dictionary of Genetic Damage* also exists as an appendix to *Genetic Catastrophe! Sneaking Doomsday?* and exists as an appendix to *Derailed Evolution*.

Derailed Evolution (2005)

All DNA is continuously in close contact with manmade and highly reactive and mutagenic/carcinogenic chemicals as well as manmade and highly reactive and mutagenic/carcinogenic radiation. This situation is undoubtedly a general genetic catastrophe of the world. Also, this situation means that human evolution is derailed.

Why Gould Was Wrong (2003) does contain *A Dictionary of Genetic Damage* as an appendix.

IV   Groups on the General Genetic Catastrophe

"Sineath has a genetic disorder called alpha-1 antitrypsin deficiency, in which a defective Alpha-1 protein can lead to lung and liver disease. He has chronic obstructive pulmonary disease and suffers emphysema, chronic bronchitis and asthma. [...] About 100,000 people nationwide [USA], or one in 3,000, are estimated to have Alpha-1, Strange said. Yet only about 5 percent of those people are diagnosed."

Postandcourier.com

"President Bush signed the bill into law Wednesday, banning insurance companies and employers from discriminating against people based on genetic test results. The law also makes it illegal for insurance companies to increase group premiums based on genetic information."

Postandcourier.com

General Genetic Catastrophe Group (A YAHOO! GROUP)
Founded: 2003, May
Messages: Ca. 11,800 (Dec. 2010)
Members: 18 (Dec. 2010)

Derailed Evolution (DE) / General Genetic Catastrophe (GGC) (A GOOGLE GROUP)
Founded: 2005, January
Messages: Ca. 10,000 (Dec. 2010)
Members: 55 (Dec. 2010)

Evolutionary Psychology Group (A YAHOO! GROUP)
Founded: 2003, November
Messages: Ca. 20,500 (Dec. 2010)
Members: 269 (Dec. 2010)

The International Association for DNA Protection (A GOOGLE GROUP)
(See letter below.)
Founded: 2010, August
Messages: Ca. 100 (Dec. 2010)
Members: 1 (Dec.2010)

The International Association for DNA Protection (A YAHOO! GROUP)
(See letter below.)
Founded: 2010, August
Messages: Ca. 100 (Dec. 2010)
Members: 1 (Dec.2010)

Dear All!

I cannot find a suitable international association for DNA protection. But we can form a new one. (NB: A non-profit association.) A good model for this association is perhaps The International Association for Food Protection.

The name of this new association could be: The International Association for DNA Protection.

The association rules must be carefully drafted. The association must operate according to its rules and keep its finances in order. The association must also follow national and international laws and common decency.

The founding of this association should be done as follows: 1. The association rules are drafted and a founding meeting is called. 2. The founding meeting convenes. 3. Another meeting is held where the rules and regulations are passed again. 4. The minutes, the founding document and a freeform application are written. 5. A homepage is founded. (And the hard work for DNA protection begins.)

Please contact me now so that we can start the founding work as soon as possible. Perhaps the founding meeting can take place in Tromso, Norway, in 2011. My email address is

n-oeij@online.no.

Best regards,

Nils K. Oeijord

V   About Nils K. Oeijord

"Although It has taken *Homo sapiens* several million years to evolve from the apes, the useful information in our DNA, has probably changed by only a few million bits. So the rate of biological evolution in humans, Stephen Hawking points out in his Life in the Universe lecture, is about a bit a year."
Whatismethaphysics.com

Nils K. Oeijord was born in Norway in 1947. A graduate of the Agricultural University of Norway, he also studied mathematics at the Universities of Trondheim and Tromsoe, in Norway as well. He is a former assistant professor of mathematics at Tromsoe College, Norway, and is the author of several scientific works in Norwegian and English. Nils K. Oeijord is the discoverer of the general genetic catastrophe (and derailed evolution) and the co-discoverer of the *Bronston heritability coefficient.* He published the first book on the general genetic catastrophe, the first book on derailed evolution, the first dictionary of genetic damage, the first dictionary of human instincts, and the first Scandinavian book on the Waldsterben. Nils K. Oeijord's books in English are *Human Instincts Explained* (2000), *A Dictionary of Human Instincts* (2001), *Human Behavior: The New Synthesis* (2001), *Genetic Catastrophe! Sneaking Doomsday?* (2002), *A Dictionary of Genetic Damage* (2003), *Why Gould Was Wrong* (2003), *The Very Basics of Tensors* (2005), *Derailed Evolution* (2005), and *Why Minus Times Minus is Plus* (2010). Oeijord is the founder of three Yahoo! Groups: *General Genetic Catastrophe Group* (2003), *Evolutionary Psychology Group* (2003), and

The International Association for DNA Protection (2010), and two Google Groups: *Derailed Evolution / General Genetic Catastrophe* (2005) and *The International Association for DNA Protection* (2010). He earned a place in *Who's Who in the World* (28th Edition). He is the founder and transitional president of *The International Association for DNA Protection* (2010).

VI   Glimpses of my experiences

"The diseases of the present have little in common with the diseases of the past save that we die of them."
Agnes Repplier

"First of all, many human diseases are influenced by, if not caused by mutations in genes."
Daniel Nathans

1. The 1940s

"First of all, many human diseases are influenced by, if not caused by mutations in genes."
Daniel Nathans

I was born May 10, 1947, in a German barracks in the very little town of Mo i Rana in Northern Norway. I grew up on a very little farm near Mo i Rana. We had a horse, a wood stove, and an oil lamp. (My mother was extremely cautious with the oil lamp to avoid carbon monoxide poisoning. I learned about dangerous chemicals from my mother.) No tractor. No cars. No electricity. I, the child, was living in a paradise. Endless exciting forests. Lakes, rivers, and creeks full of life. Mountains with snowy tops. Glaciers. Eagles ruled the sky. Fantastic springs. Summers full of butterflies and singing birds. Yellow and red jaw-dropping autumns. The first snow. Heaven on Earth. Beautiful winters. Unbelievable animals in the wilderness. But things would change.

In the 1940s I did not know, of course, that radioactive emissions from Sellafield (formerly known as Windscale) began in 1947, the year I was born. Sellafield is a nuclear fuel processing plant in UK. See more about Sellafield in the next section.

2. The 1950s

"I believe that pipe smoking contributes to a somewhat calm and objective judgment in all human affairs." (Albert Einstein. Statement upon joining the Montreal Pipe Smokers Club in 1950.)

The late 1940s / early 1950s is my measuring stick. The zero point of my measuring stick is the time of the horses before the first cars and the first tractors appeared in the earthly paradise of my childhood. If we do not have a measuring stick with a zero point on it, we cannot measure and understand our environments and our surroundings.

In 1946 the Norwegian parliament decided to build the state owned Norwegian Iron Works, Inc. in Mo i Rana. The iron production started in 1955, and lasted until 1988. The pollution was enormous. Wild birds were colored red. The iron works was placed in the town itself! The workers as well as the general population did inhale enormous quantities of air pollution. They were told that the pollution was harmless, and that enzymes in the lungs did totally remove the dust and gases from the lungs. In general, protection equipment was not used. The workers smoked heavily while standing in the dust and gas clouds. The food in the canteen consisted of black coffee, sugar lumps, and Danish pastry. The state paid doctors told them that smoking was good for them, especially if you were an asthmatic. Science did not understand that environmental poisons did exist at all. Now, just after WW II, gene damage and genetic diseases were taboos. Data and results from earlier genetic research were destroyed. The crazy orders to do these things probably came from the socialist government who paid the scientists. And most scientists seemingly had no thoughts or will of their own.

In the 1950s the cars came to Mo i Rana. Walking along a country road you smelled and inhaled the terrible diesel exhaust or gasoline exhaust when a car drove by. I instinctively understood that all these changes were basically and fundamentally bad for us humans as well as for the wilderness and all the wildlife. Paradise lost.

In the 1950s (and 1960s) state paid dentists attacked the teeth of all children and teenagers of Northern Norway and the rest of the country. They drilled a lot of holes in flawless teeth in every individual and filled the holes with mercury and other heavy metals. In general heavy metals are mutagenic. Mercury atoms inside a DNA molecule cause single strand breaks and are, of course, a genetic catastrophe. Mercury is mutagenic, carcinogenic, embryotoxic, teratogenic, has several other pathological effects, and causes a reduction in immune function.

In the late 1950s state paid dentists destroyed my flawless teeth, and poisoned my body with mercury. Since then my brain and the rest of the central nervous system were damaged in several ways. My wife experienced exactly the same story back then. Now she has already suffered several strokes in addition to a life with migraine disease.

By the way, today mercury in seafood is a worldwide health problem. In the USA, the Food and Drug Administration warns that pregnant women, nursing mothers, women who might become pregnant, and children should not eat swordfish, shark, tilefish, and king mackerel because of their high mercury content. The Food and Drug Administration also warns women and children to limit their consumption of tuna. Remember: Mercury is mutagenic, and people do not know that. *People without knowledge elect politicians without knowledge in a world full of uninterested scientists.*

When I was a little child my parents and I visited a house where a man was smoking heavily. My family did not smoke, and this was the first time I experienced tobacco smoke. I still remember clearly that the room was totally filled with grey tobacco smoke. I, the little boy, played with toys on the floor, but I remember that I looked repeatedly at the strange smoke. Several years later the supersmoker became the father of a girl with Down's syndrome. Later in life I learned from my own observations that heavy smoking is linked to Down's syndrome and other genetic diseases. Repeatedly I discovered new links between manmade pollution and genetically determined damage to human health. But where was science? My experiences lead me to lose respect for scientists and politicians, and even the general population.

When I, for example, said that I had good reason to believe that smoking can cause Down's syndrome, I was met with laughter, and laughter only.

In the 1950s we Norwegians did not know about the state-owned Sellafield (see above). But our seafood did contain Technetium-99, Cesium-137, Zink-65, and a large number of other manmade radioactive isotopes created by Sellafield, and unknown to natural nature. These mutagenic manmade isotopes are environmental poisons, accumulating in living organisms. All human cells in Norway are now housing these radioactive devils, continuously bombarding DNA molecules until the end of humankind. The tragic Windscale fire (1957) released some 750 terabecquerels of radioactive materials into the environment of the world. (A becquerel (Bq) is a unit of radioactive measurement, representing one nuclear disintegration per second. Tera = 1,000,000,000,000.) Totally there have been more than 20 accidents in Sellafield involving radioactive pollution. And there have been deliberate discharges to the atmosphere of plutonium, etc. Nuclear waste from Sellafield is found in supermarket salmon. There is plutonium inn all children's teeth. To the end of humankind. Science don't really care. The message from mainstream Science is: It's harmless. See more on humanmade radioactive mutagenic pollution in later sections. On radioactive pollution from Sellafield in the UK, see, for example, www.lakestay.co.uk/hot.htm, and be shocked. On low-level pollution exposure, see, for example, www.ratical.org/radiation/inetSeries/NIDcell.html, and be shocked.

A glimpse from 1956 (quote):

"HAZARDS OF RADIATION
MEDICAL RESEARCH COUNCIL'S REPORT

The main conclusion of the Medical Research Council's report on nuclear and allied radiations,' published this week, is that "adequate justification should be required for the employment of any source of ionizing radiation on however small a scale." This follows a review of the effects of radiation from such sources as nuclear weapons, medical radiography, X-ray apparatus for shoe-fitting, luminous watches, television sets, cosmic rays, isotopes, and industrial X-ray machines.

1418 JUNE 16, 1956 HAZARDS OF RADIATION MEDICAL JOURNAL"

(Unquote.)

When I grew up we had vinyl (PVC) tiles on the kitchen floor, we got drinking water through PVC pipes, we wore raincoats made of PVC, and so on. But I was too young to know. Well, the tiles are removed long ago. They were finally burned? But the world is still using PVC, and is still burning PVC. They are producing dioxins. Where is science? PVC was first created in 1872, and was patented in 1913. Plasticized PVC was invented in 1926, in the US, of course. Burning PVC, and even PVC production, produces catastrophic dioxin pollution. (Please use natural materials only.) Dioxins are among the most toxic and mutagenic chemicals ever produced, and they don't go away. But, in the 1950s, I was too young to know. And certainly, the scientific community, was too dull to know. The dioxin pollution of the world is *alone* a global and local general genetic catastrophe. *Burning organic substances containing chlorine produces dioxins.* Dioxin pollution from pulp and paper production is due to the use of many million tons of chlorine per year to bleach wood pulp white. Paper, of course, contains dioxins, all mother's milk contains dioxins, all water contains dioxins, all everything contains dioxins. Believe it or not, scientists, in general, are explaining away the whole dioxin catastrophe. They are too "young" to know.

3. The 1960s

"Alzheimer's, Parkinson's, brain and spinal cord disorders, diabetes, cancer, at least 58 diseases [...] that touch every family in America and in the World." Rosa DeLauro

In the 1960s US researchers proved the relations between tobacco smoking and lung cancer. But the workers of the Iron Works and other state owned companies were mostly socialists and communists and did not want to listen to US scientists.

The state owned Norwegian Coking Plant Inc. in Mo i Rana started production in 1964 and was closed down in 1988. This Coking Plant was placed in the little town itself! In the plant dust, gases, and heat were terrible. Workers fainted because of exposure to CO. The exposure to mutagenic and carcinogenic PAHs during one day in the worst places of the plant equated PAHs from smoking some 6,000 cigarettes. The floors were washed with mutagenic and carcinogenic benzene! It was said that all green plants in the administration building died overnight due to poisoning. And, of course, the state owned Iron Works and the state owned Coking Plant were economic and environmental disasters. But science and scientists were silent.

Rachel Carson, a US biologist, ecologist, and author, published *Silent Spring* in 1962 and thereby started the environmental movement of the world. *Silent Spring* was later generally listed as the best non-fiction book of the 20th century, and was recently named one of the 25 greatest science books of all time by the editors of *Discover Magazine*. *Silent Spring* took Carson four years to complete. The book contained a long list of experts who had read and approved the manuscript. In 1972 DDT was banned, but the whole world was already heavily polluted with mutagenic and carcinogenic DDT. (Most scientists even today conclude that DDT is not mutagenic and / or carcinogenic, but I believe DDT reacts with other chemicals in the general pollution, so that terrible things happen to our genes and chromosomes, due to DDT pollution.) Rachel Carson died in 1964 from breast cancer, and so became a victim of the general genetic catastrophe. Many scientists were claiming that Carson was a hysterical woman. Science in general was pretty silent. Many chemists and other scientists attacked Carson. President John F. Kennedy ordered a science committee to examine the issues of *Silent Spring*. The committee vindicated both *Silent Spring* and its author.

I read *Silent Spring* in 1963. I remember that the people of Northern Norway – from high school students to university people – did not take the book seriously. I was disappointed. I learned that the general population and the science people were pretty lame. They simply lacked certain talents. And I did understand why it was Rachel Carson versus the silent scientists. By the way: Why should the leftists, who were dominating the Press, agree with Rachel? Their prophet Karl Marx said that a paradise on Earth was coming. Moreover, their religion was Historical Human-made Progress based on Science and Science-based Technology in a Godless Universe. They could not easily believe in Rachel.

I discovered the general genetic catastrophe in the same way as Rachel Carson discovered the worldwide general catastrophe of environmental poisons. The discovery of the GGC did not come from research findings in the traditional sense, but from total analytical overview, rather than one or more narrow analysis.

When I was in the Norwegian military 1967 – 1968 and seven short periods in the 1970s and 1980s, I experienced how military personnel were victims of dangerous and illegal radar radiation. I was a sergeant and did my best to protect people from being victims of radiation from the radars. I was close to be arrested by the MP. But the MP understood very fast that I was deadly serious. In the military I learned that science and society did not understand the real dangers of mutagenic pollution (chemicals and radiation).

4. The 1970s

"When it comes to adolescent eating disorders, it is usually the parent's responsibility to seek help. The girl with the eating disorder is often the last to know she is ill."
Anna Pacheco

I finished my education in plant production, genetics, plant breeding, etc, in 1973. I was shocked to learn about the heavy use of pesticides, and heavy use of synthetic fertilizers containing heavy metals, and other mutagenic substances.

Modern science-based agriculture is mutagenic. Mutagenic diesel exhaust particles (see later) are covering the crop fields, and mutagenic (directly or indirectly) pesticides are sprayed over the crops. Although nitrous oxide ($N2O$) is mutagenic, it is widely used in synthetic nitrogen fertilizers and anaesthetics. Phosphate minerals in phosphate fertilizers contain the following mutagenic heavy metals (and environmental poisons!): Cadmium, lead, arsenic, chromium, nickel, mercury, uranium, polonium-210. By the way: Tobacco smoke contains highly radioactive polonium-210 (from phosphate fertilizers). New study links mercury to dementia and Alzheimer's disease.

The combination nitrite + bisphenol A (BPA) is shown to be mutagenic. (Nitrite is well known in modern agriculture.) Study shows that 93 (sic) percent of the human population has some BPA in their bodies.

In the years 1973 – 1976 I worked as a researcher at an agricultural experimental station in Western Norway. While I worked there I started collecting seeds of old and rare Norwegian varieties of food plants. I collected totally 120 varieties and established Oyjord Gene Bank and published totally 8 issues of *Gene Bank News*. This little private gene bank inspired the Nordic countries to speed up the creation of Nordic Gene Bank. I participated in a founding meeting in Norway. Later I sold my seed collection to Nordic Gene Bank and Oyjord Gene Bank became history. During my gene bank work I learned that *the Norwegian state paid plant breeders had exterminated all old Norwegian land varieties of cereals and even exterminated the first and second generations of the new Norwegian scientifically bred varieties. An enormous genetic catastrophe. Similar genetic catastrophes happened all over the world thanks to state driven scientific plant breeding.*

When I moved back to my home district Rana in 1976 I was shocked when I learned that the best soil in the whole district was planned to be destroyed by the building of some 1,000 buildings and a lot of roads, etc. I immediately started a soil protection campaign which saved the soil resources. The Norwegian government listened to me, and decided to stop and delete the terrible destruction plans. On this land, then, a lot of local wild plant genes of future value for natural pastures were automatically saved. A little local genetic catastrophe was avoided.

5. The 1980s

"I've had this problem since I was in my 20s. They don't call it manic depression anymore. They call it a bipolar disorder, and I'm a Type 2."
Ned Beatty

Forest dieback (Waldsterben in German) was broadly discussed in the 1980s. I researched the forests of Northern Norway and published two books (1985 and 1987) on the damage to Norway spruce (Picea abies). I contacted Norwegian newspapers, gave newspaper interviews and radio interviews, published articles in newspapers, gave lectures, etc, to inform the public and the politicians about the damage to our forests. The Norwegian government agreed with me that the situation was dangerous, but the Norwegian state paid forest researchers totally disagreed with me. They claimed that the Norwegian forests were completely undamaged. They also claimed that the forests in Europe as a whole were undamaged. We were extremely few people in Europe who understood the dangers and causes of the Waldsterben, and worked hard to spread information on the situation. I wrote the first Nordic book on Waldsterben (written 1984 and published 1985). In Norway I now remember only two names of really important "activists", including myself. And I remember one in Finland, one in Sweden, one in Denmark, and three in Western Germany.

Acid rain killing fresh water fish in rivers and lakes in parts of Scandinavia was a part of this nightmare. Luckily the European politicians did listen to the "activists" and did not listen to the state paid research people who explained away the whole thing. The use of brown coal (several percent sulphur) and high sulphur oil were drastically reduced, stone coal mines were closed, natural gas was used instead of oil for heating and cooking, emissions from power plants were cleaned up, an so on. The problems of acid rain were dramatically reduced, and the forests of Europe did not totally collapse. And many fresh water fish populations survived. Today, some 25 years later, I now do understand that I did my part of the work necessary to avoid a full general genetic catastrophe in the wild nature of Europe.

But what exactly is the situation today? Well, according to German statistics, in 2009, in Germany 2/3 of all trees are sick. And in two of the most important tree species 80 percent of the trees are sick. Today, Science is neglecting Waldsterben.

In the 1980s I learned that the Norwegian scientific community did not believe that pollution (CO_2, etc.) causes a warmer Earth. They even did not believe that the ozone hole(s) did exist. They ridiculed the US research results on the thinning of the ozone layer. Even in the 1990s Norwegian teachers did not believe in manmade global warming, I learned. (Norwegian teachers and researchers are mostly leftists. By the way: Norwegian leftists strongly dislike the US and all talk about genetic things; see Chapter I.)

In the 1980s (and earlier and later) Norwegian university laboratories were polluting students and producing a lot of cancer victims. At the University of Oslo the water taps contained mutagenic and carcinogenic cadmium and polluted an unlimited number of people for a long period of time. In

Norway the drinking water was, and still (2010) is, uncleaned rain water. And not only that, but mutagenic and carcinogenic chlorine is added to the water. The tap water (drinking water) smells and tastes like a swimming pool. And, of course, often the water tastes and smells general pollution from near and far. The science-backed authorities tell us that the water is cleaned, but the water is not cleaned at all. On the contrary, the water is polluted with chlorine to kill microorganisms.

In the 1980s *black snow* (full of soot) became common everywhere in Norway (and the world). Soot is mutagenic and carcinogenic. Also, *red snow* became common. Red snow is caused by bacteria feeding on the pollution in the snow.

Science, seemingly, did not to see the wider implications of these dramatic new facts. Not only that, but science seemed pretty uninterested in the whole thing. No wonder I generally do not trust scientists.

In 1986 the state owned Chernobyl nuclear power plant in the Soviet Union exploded. The radioactive release of the Chernobyl is calculated to equal some 400 Hiroshima bombs. But Chernobyl is much more damaging than that because the isotopes from Chernobyl are more long lived. The Norwegian government was almost totally passive after Chernobyl, and Norwegian scientists said that the radioactive pollution from Chernobyl was harmless. But reindeers, cheep, brown trout, mushrooms, etc. rapidly accumulated enormous amounts of radioactive isotopes, and were no longer food for humans. In reindeers in central Norway the radioactive pollution wase as high as some

140,000 Bq per kilogram. (Average was some 70,000 Bq per kilogram.) After Chernobyl, the Norwegian safety limit was set to 6,000 Bq per kilo reindeer meat. The radioactivity did not go away. Example: The level of Cesium-137 from Chernobyl reached 7,000 Bq per kilogram in cheep in 2006. The farmers must reduce the level of radioactivity in cheep by giving them non-contaminated food for a month before slaughter. However, *all safety limits are politically established to protect agricultural production and the reputation of science.*

In reality there is a continuous release of radioactive material from nuclear power plants into the environments of the world. Moreover, radiation released from fossil fuel power plants around the world are continuous and mostly of long lived isotopes, so one needs to total up releases for the past some 300 years or so. The general population of Norway does not know these things. More: In produce water from the Norwegian oil and gas production, elevated levels of Radon-226 and Radon-228 are discharged to the sea. So far no study has assessed the radiological effects on seafood and humans in connection with radionuclide discharges to the sea. By the way: Nuclear power plants produce (and discharge to the environment) about 500 types of radioactive atoms (unstable nuclides).

The Novaja Zemlja Test Site (not far from Northern Norway) conducted 224 nuclear explosions 1955 – 1990. Iodine-131 from these nuclear explosions causes thyroid cancer in Norwegians, but the Norwegian government and Norwegian scientists are pretty uninterested, and , of course, *general genetic damage caused by radioactive pollution is not discussed at all.*

Over 2000 nuclear explosions have been conducted around the world. And the list of nuclear accidents is shockingly long. See Wikipedia: *List of civilian nuclear accidents*, and *List of military nuclear accidents*. See also: *List of nuclear accidents* at www.fact-index.com.

The bulk of all plutonium is manmade, because plutonium is short-lived on a geological timescale. Totally and shockingly some 7700 kilogram plutonium have been released into the atmosphere. At Sellafield some 500 kilograms of plutonium and 40 other radionuclides have been released into the sea. Plutonium from Sellafield is found in the Arctic. People living near Sellafield have thousands of times more plutonium in their bodies than other people. If you do google "there is plutonium in" you will (hopefully) learn a lot. See more on Sellafield pollution in later sections.

The following report proves that in the 1980s science was unable to understand the problems of Hiroshima and Nagasaki.

The copy (quote) below is copied from:

http://www.ornl.gov/sci/techresources/Human_Genome/project/alta.shtml

(NB: The quote below is from an article published December 1984.)

Quote:

"Existing methods had failed to detect an anticipated increase in mutations among the more than 12,000 children of Hirsohima-Nagasaki survivors (whose parents received an average 43 rad). Calculations showed that to measure a 30% increase in the mutation rate, roughly what would be expected from the average dose, one would have to examine $4.5 \times 10^{(10)}$ bp in the children, and 4 to 5 times more in the parents (Delahanty, 1986). In fact, the DNA methods were at least an order of magnitude short of being able to detect the expected impact from atomic bomb exposure among survivors; they could only detect differences expected from radiation exposure well above the lethal dose (and hence not measurable). The question was whether there were new technical means that would get around the problems. The answer was no, but the process of thinking about it forced many novel ideas to the surface."

Unquote.

6. The 1990s

GGC example: hydrocephalus:

"Mutations in L1CAM, the gene encoding the L1 neuronal cell adhesion molecule, lead to an X-linked trait characterized by one or more of the symptoms of hydrocephalus, adducted thumbs, agenesis or hypoplasia of corpus callosum, spastic paraplegia, and mental retardation (L1-disease). We screened 153 cases with prenatally or clinically suspected X-chromosomal hydrocephalus for L1CAM mutations by SSCP analysis of the 28 coding exons and regulatory elements in the 5'-untranslated region of the gene. Forty-six pathogenic mutations were found (30.1% detection rate), the majority consisting of nonsense, frameshift, and splice site mutations. In eight cases, segregation analysis disclosed recent de novo mutations. Statistical analysis of the data indicates a significant effect on mutation detection rate of (i) family history, (ii) number of L1-disease typical clinical findings, and (iii) presence or absence of signs not typically associated with L1CAM-disease. Whereas mutation detection rate was 74.2% for patients with at least two additional cases in the family, only 16 mutations were found in the 102 cases with negative family history (15.7% detection rate). **Our data suggest a higher than previously assumed contribution of L1CAM mutations in the pathogenesis of the heterogeneous group of congenital hydrocephalus**." [Bold type for emphasis made by NKO.]

Am. J. Med. Genet. 92:40–46, 2000. © 2000 Wiley-Liss, Inc.

Mo Industrial Park (1988 -) in Mo i Rana houses (2010) about 135 companies, many of them producing metal alloys. The park is releasing tons of genotoxic mercury pollution into the environment, that is hundreds of kilograms mercury per year. The Norwegian government and the workers allow them to do so, even though they know that no fish in Norway is free of mutagenic mercury pollution. The total mercury pollution of the world is (year 2000) some 2000 tons per year. (In 1980 some 6000 tons per year.) The mercury pollution of the world is a general genetic catastrophe of its own. Reading the following article by a researcher, we understand better the enormity of the mercury catastrophe. The article below is quoted from www.markschauss.com.

Quote:

"September 27th, 2006

[..] Let's put this into prospective. If we think about the fact that about 200 milligrams are lethal to humans (goes up and down dependent on a number of issues like genetics and environment as well as the type of mercury) and if we release say a mid-range number of 5,500 tons of mercury into the atmosphere, what kind of number are we really talking about?

5,500 tons of mercury translates to 24,250,549 kilograms which then breaks down to 24,250,848,840,337 milligrams. This is 24 and a quarter trillion milligrams. Now if we estimate that the total human population is about 6.5 billion, this means that we are pumping 3791 milligrams of mercury per human being into the environment. That is 18 times the lethal dose. Now of course,most peopledon't get exposed to that much mercury but we also have to understand that this is a bioaccumulative toxin. It keeps building up in our systems over the years.

Let's look at another reality. If we take a guess that we are only going to come in contact with 1% of that mercury, how long before we hit the lethal dose? In only 5.4 years we will be exposed to the lethal dose. Of course, we do excrete mercury as we get exposed, some better than others. Let's now suppose that we excrete 75% of that mercury (it's probably less). We would then have accumulated the lethal dose in 21.44 years. If we excrete 90% then we would hit that level in 54 years. Frightening isn't it?"

Unquote.

In the 1990s I became fully aware of the general genetic catastrophe. I wrote the book *Human Instincts Explained* in 1999. It was published in 2000. In this book the mechanism of the general genetic catastrophe is explained in detail. (See more about my books and groups in chapter III and chapter IV, respectively.)

Since the late 1990s an unexplained and dramatic extinction of species has taken place in Europe, especially the UK. According to the researchers, climate change cannot explain what is going on. And note that climate change is caused by human made pollution. Of course, this is a general genetic catastrophe, but is it a general genetic catastrophe caused by manmade general mutagenic pollution? We do not know.

In the 1990s the scope and the scale of the destruction of life on Earth became well known among the general population of the world.

Shocking new statistics on thousands of genetic diseases appeared in the 1990s. Shocking new statistics on environmental poisons in humans, animals, food, milk bottles, toys, freshwater, seawater, etc., appeared. Yes, to me it was shocking. But, as far as I learned, Science, in general, was not shocked and not alarmed. A scientist (leftist!) called me an alarmist in the bad sense of the word. He had no further comments to make! Obviously, in general, Science was (and still is) out of touch with this fundamental and ultimate catastrophe of the world. By the way, as I see it, *catastrophe* is too weak a word.

Now back to radioactive pollution from Sellafield. In the 1990s, along the Norwegian coast, the highest measured Technetium-99 concentrations in old parts of the brown seaweed were almost 1000 Bq per kilogram dry weight. The researchers measured Technetium-99 concentrations in other sea organisms as well. But they did not measure Technetium-99 pollution in seafood such as economically important fish species. It is not difficult to understand why. And what about the other radionucleides? *This is not science! Science is neglecting and ignoring the general genetic catastrophe of mutagenic radioactive pollution..*

Dr. Rosalie Bertell Ph D (1929 -) is an internationally recognized expert in the field of radiation. She is the author of two extremely important books: *No Immediate Danger. Prognosis for a Radioactive Earth* and *Planet Earth. The Latest Weapon of War.*

I found, on the internet, an article by Rosalie Bertell published in 1986, the year of the Chernobyl explosion and fire. I hope Dr. Rosalie Bertell does accept that I below do quote the whole article. Dr. Rosalie Bertell's article is extremely informative. Here comes a full length copy, from the internet:

Quote:

"The following, "No Immediate danger? Prognosis for a radioactive earth," appeared in *Women and sustainable development: a report from Women's forum in Bergen, Norway, 14-15 May 1990*, published by the Center for Information on Women and Development, Oslo, Norway, 1990, pp. 18-21.

Dr. Rosalie Bertell has a doctorate in Biometry, which is the use of mathematics to understand and predict biolological processes, for example in cancer research. For more than twenty years she has worked with health-oriented environmental protection with special emphasis on the consequences of radioactive radiation. She is the President of the International Institute of Concern for Public Health in Toronto, Canada, and one of the founders of the International Commission for Health Professionals in Geneva. This Commission works with health personnel to secure human rights.

Dr. Bertell has been a consultant for the US Nuclear Regulatory Commission and the US Environmental Protection Agency. In her book "No immediate danger, Prognosis for a radioactive earth,"(Women's Press, London 1985) she sums up the damaging effects of low-level radiation, the genetic damage to human beings as well as the effects of the increasing background radiation from nuclear power plants and the nuclear weapons industry.

Dr. Rosalie Bertell:

No immediate danger? Prognosis for a radioactive earth

The operating mode in our Society is called risk-benefit planning. Everything is done on the basis of risks and benefits. Risks are life and health--dying of cancer, having a deformed child. The benefit side is to make money or to gain political or social power. The bad news is that the people who make these rational trade-offs for us, are the same people as get the benefits.

I will talk about sustainable development in a very different way than what they do at the official Conference. I would compare this to your own personal experience: If you have gotten beyond youth and good health and suddenly have to confront chronic illness, you begin to realize what it is to sustain what you are doing. The kind of work you do, the amount of sleep, the amount of food, the amount of new undertakings are very much limited by your personal energy and your personal ability to sustain it.

In the same sense the human race has limitations. We act as if that part is all automatic, but that is not true.

Damage the future generations

Another unconscious assumption is that as long as we make sure there are enough resources around in the future, there is no problem for the future generations--they can just go ahead and do what they want. In fact we talk about "the freedom of choice for the future generations."

I want to talk about something called mild mutations which is a very subtle undermining of the genepool. It is not talked about, it is not measured, but it is occurring. What you do is to create a next generation that is physically less able to cope with hazardous material than their parents were. If you do two things at once: you mildly damage the next generation--genetic damage--and you increase the hazards in the environment, then you can do this for two or three generations and you are finished. People will be physically less able to cope, and they will have more to cope with.

We are also talking about physical damage to the brain, inability to think as clearly and as well as previous generations. With aboveground weapon testing there was a decrease in general intelligence quotient as measured by standardized tests. They are starting to admit it now even in official publications. In a recent publication from the US Academy of Science it is admitted that exposure of the fetus to radiation between the eighth and the fifteenth week lowers mental ability, causes mental retardation.

I am going to use radiation as the basic pollutant, because it is so all pervading.

Explosion in living cells

Non-ionizing radiation is the use of long wavelengths for radio, heating, microwaves, infra-red etc., while penetrating ionizing radiation which occurs in gamma-ray, X-ray and cosmic-ray radiation is the radiation I will here refer to here. When a material is radioactive it means it periodically has an explosion on a microscopic level. When it explodes, it gives out energy in such a way that it can ionize--penetrate. Take for example just one atom of plutonium in a lung tissue. In exploding it shoots out particles of energy through living cells. As you know a cell is not empty, but a living system filled with different types of matter with separate jobs to do in the body. We can not feel anything of this explosion on cellular level. But it will do damage.

There is no such thing as a radiation exposure that will not do damage. There is a hundred per cent possibility that there will be damage to cells. The next question is: which damage do you care about?

The damage which is apt to cause most trouble in a whole system like a human being, is the damage that hits the nucleus of the cell. Because inside the nucleus is the chromosome material that carries the template of what the cell does. If you change that, you change what the cell produces. If you change one cell, and it is still able to produce, it makes two cells with damaged chromosomes which can cause exponential growth of cells that are not going to do the right thing. An example of an illness that results from this is adult diabetes where the person has sufficient insulin, but it doesn't work to bring down the blood sugar. It is mutated.

Another example is allergy. You may get allergy as an adult--then you have stopped producing antibodies. Over time we build up things like inability to digest food, we don't recover as quickly after illness, we get hormone problems and so on. All this comes from this kind of changes and mistakes in the chromosome material.

Waste precious resources

The one type of damage talked about the most, is when you upset a cell's resting mechanism. Then you get a cancer or a tumor. Normally a cell reproduces and then rests. If you eliminate that rest, the cell just reproduces all the time and you get a lot of cells in one place, a tumor.

You know it takes only one ovum and one sperm to make a baby, and the DNA in that cell contains all the information on how to make a normal baby. If you start destroying that DNA, you get a deformed child or a sick child.

We always have to remember that the future generations on this planet are not nebulous, we are right now carring them in our bodies, they don't come from out of space! They come from the sperm and the ovum that are right now living in the bodies of people living on this planet. If we destroy that, we have no way of putting it back together again.

To sum up: Mild mutation is damage to the genetic material which shows up in the children as either obvious disability or asthma, allergy, immune deficiency, childhood cancer, etc. These are very real and are occurring. We can increase the rate of the damage, and we have done just that. This is probably the most insane way to waste the precious resources of the world. Whatever I say about humans also is true in the animals and the plants that produce our food. We are killing ourselves with our survival strategy. We think we are saving ourselves, but instead we are undermining our ability to survive.

Not only cancer

For many years we have known that radiation causes damage to human cells, and we have heard about genetic damage and cancers. But already in the fifties medical radiologists started looking at their own first causes of death and discovered that not only were they dying at a higher rate from cancer, but that they also were dying at an higher rate from everything else: cardiovascular, renal disease and other chronic illnesses. I am trying to show you the breadth of what radiation does. We receive so much propaganda saying that if you are exposed to radiation you have a chance of getting cancer, and if you don't get cancer you don't get anything. But to repeat: there is a hundred per cent probability of cellular damage when you are exposed to radiation. The body can repair some, but there is always residual damage. Whether or not you care about it depends on how severe it is and what you take as a starting point. Right now they are counting only fatal cancer. You are told that low doses of radiation can't be measured.

Analysis of cancer deaths in children in the US during the aboveground nuclear testing period from the late forties to the sixties shows a startling increase in childhood cancers between the age of zero and five. They respond to radiation doses in utero. The rates are now coming down, but have not yet returned to the level before the testing started. Downwind of a test-sight is even worse. We are talking here about hundreds of babies. Anything that can cause death by leukemia is going to cause a lot of other effects in children. For every child with leukemia, about 100 others have changes in their blood.

The most remarkable thing was that no one complained or protested. But they didn't know. Each woman and man who lost a child, thought it was a personal tragedy and didn't connect it with the weapon testing. But the analysis states that two out of three child deaths were unnecessary.

Deaths of newborn babies

There are other public health dangers from "normal" radiation. I am talking about normally operating nuclear power plants, not accidents. We have analyzed the death rate of babies of less than 2,500 grams near nuclear power plants in the state of Wisconsin. Over the first five years of normal operation of the nuclear reactors there were a hundred excess infant deaths. Again nobody was complaining or marching in the streets-- because each one was a "family tragedy." In fact I went back and visited the hospitals and the doctors knew that their child death rate was going up. But they did not look outside the hospital for reasons. They were bringing in experts to look at their intensive infant care unit, and they were sending their nurses out to learn how to care for babies. So the nurses were being blamed! If it is not the nurses, it is the mothers to be blamed when there is a deformed child.

But these deaths don't count in the ordinary risk/benefit calculations, because these babies don't die of cancer. So we are into an insane system of what counts. There are other infants who survive, but who live with disabilities.

If we look at miscarriages the same thing holds true. You damage the sperm cells and ovums and a large number of babies are lost. I worked with the pollution from the nuclear plants at the Love Canal in the States, the only area to be declared a national disaster area for pollution. We had already evacuated a hundred families in the center of the pollution. During the following winter they were deciding whether or not to evacuate the people in the outer circle. Of ten babies born in that period in this outer ring only one was normal. And that was the pregnancies that came to term, there were a lot of miscarriages.

This should show the vulnerability of the human race. If we don't survive none of the rest of this planet is going to be worth anything.

Marshall Islands: Experiment on human beings

Let me tell you about the Marshall Islands. On 1 March 1954 they set off the first hydrogen bomb on the Bikini Islands. The bomb represented 1,000 Hiroshima bombs! The island of Rongelap got the fallout of the bomb. This was actually a very deliberate experiment on human beings. That bomb was set off and the people downwind were not evacuated. They wanted quite deliberately to find out what would happen if the population "only" got the fallout and no blast effect, like they did in Hiroshima. They finally evacuated the people after 72 hours and gave them medical care. Three years later they said the island was now inhabitable, it fitted within the "permissible level of radiation." They not only moved back the "fallout" population, but sent a matched control group back to find out how these people would survive the residual radiation. They were sent to live on a contaminated land, which was within "the permissible level." By the way, this level is the same one that you used in Norway for the Chernobyl fallout. An analysis of the children from the control group shows that the white blood cells were destroyed. That might not seem important to you, but the white blood cells are the ones that help you fight diseases. The children had epidemics of polio, lepracy and tuberculosis. Epidemic diseases occur everywhere where there have been radiation accidents.

I have now moved from what causes cancer to what disrupts your blood system. If we based our regulations on what disrupts your blood system, the permissible level of radiation would drop dramatically. How many becquerels in the reindeer would be permissible if it was based on whether or not your blood was affected? Then that number would drop dramatically. What we are saying now is that: well, the Sami people are eating it, the rest of us can get something else to eat, so we don't bother.

We leave the regulation at a level that kills people, that destroys pregnancies.

Evidence: the Kerala Community

The scientists of the establishment claim that we don't have any evidence that humans suffer genetic effects from radiation. This is based on the fact that they didn't see any genetic effects after Hiroshima.

But there are large populations that have been exposed to radiation over generations that they could look at if they wanted. A recently concluded health survey reveals that more disabled and sub-normal children are born to mothers exposed to higher natural background radiation. There is an area an in Kerala in India, where there is naturally occurring thorium monazite sand, a kind of black sand. There are 44,000 people living there, many for generations. Over the last two years we have collected information on illness among the families living on this radioactive sand compared with families living on natural sand in the same area.

What we found on the radioactive soil was four times the expected level of Down's Syndrome or mongoloid children. Also mental retardation, epilepsy, congenital blindness and deafness, cleft lip and cleft palate, skeletal abnormalities and childless couples.

When I hear about family planning, given what I know about the Marshall Islanders, the downwinders of the Nevada Test Site, and the Kerala Community, I feel like fighting for the right to have children and the right to have normal children. Because this is taken away - also in a much more subtle way. The University of Florida has measured sperm rates in men for the last 25 years in the US. It used to be 1 in 25 males in the US who were infertile. Now it is one in five. 20 per cent of the men in the US are now unable to have children. This is another human right that is being violated.

What can we do?

How can we get out of this situation? I started by saying there is a risk/benefit trade-off. One of the simplest, but most resisted things to do, is to separate out the function of calculating the risks from that of calculating the benefits. You should strengthen your public health sector. The doctors don't know what the risks of radiation are. They use statistics produced by the nuclear industry to tell what the risks are. Your Health Department is still dealing with the plague, they are not dealing with environmental health problems--and they have to, or else we will not survive. Focus on risks you care about, like long term chronic diseases, like undermining immune systems, like producing disabled children and miscarriages.

We don't have to wait until we have an epidemic of cancer deaths. We can move way back and look at more sensitive health indicators--and stop before we get to an ecological disaster.

At some point or other if we survive, there's going to have to be a massive non-cooperation with our society which is producing death. And if we are ever to break out of the militaristic society that we live in--and that *is* what I think is our basic aim, because that's what distorting every- thing—it's going to have to be through an across-the-board non-cooperation effort

-- Dr. Rosalie Bertell, Vancouver, 1986."

Unquote.

See also:
Victims Of The Nuclear Age at www.ratical.org/radiation/
NAvictims.html.

Above it was much Rosalie Bertell. But it was not too much Rosalie Bertell. However, that does not mean that I do agree one hundred percent with Rosalie Bertell. She does not, for example, fully research (quantitatively) manmade mutagenic radioactive pollution versus manmade mutagenic chemical pollution. Another example: She does not compare objectively and quantitatively manmade depleted uranium radiation and natural uranium radiation. And she does not fully quantitatively compare manmade depleted uranium pollution and pollution from Chernobyl. Therefore she is easily misused for leftist anti-US propaganda. Nevertheless, sister Dr. Rosalie Bertell Ph D is my science hero. She knew early on that she would become a nun. She has been nominated, en masse, for the Nobel Peace Prize.

"We have to be part of something larger than ourselves, because our dreams are often bigger than our lifetimes. Religion has a profound effect on our staying power." Rosalie Bertell.

The Mo Industrial Park, in Mo i Rana, Norway, mentioned above, has experienced the problems with radioactive and chemical contamination in metals and other products. However, radioactive and chemical contamination in many kinds of products and materials are pretty common today. Here comes a wake-up call:

The copy (quote) below is from:

http://ksparth.blogspot.com/2009/02/radioactive-contamination-in-steel-wake.html

Quote:

"Radioactive contamination in steel: a wake-up call

Regulatory authorities have identified Indian steel products contaminated with cobalt-60 in the U.S., Germany, France and Sweden. The events occurred at disturbingly high rates. "Overall, 123 shipments of contaminated goods have been denied entry to U.S. ports since screening began in 2003, according to Homeland Security data. Of those, 67 originated in India, 23 came from China and 20 were from Canada" (The Los Angeles Times, November 13, 2008). We cannot ignore this wake-up call. 150 incidents In the last three years, out of the 500 incidents, involving uncontrolled radiation sources, which the International Atomic Energy Agency (IAEA) came to know, 150 were related to sources found in scrap metal or contaminated goods. The finding in Germany is attracting more attention. On August 19, last year customs officers identified a container of contaminated stainless steel bars from India on way to Russia. They ordered that the container be put back on the ship immediately and be sent back to India (SPIEGEL ONLINE, February 16). There were several such findings later; The German Environment Ministry received 19 findings which included radioactive bars, steel cables, chippings and valve housings from 12 states. SPIEGEL reported that a total of 150 tons of contaminated steel has been seized. Some of it, about 85 tons, according to a reliable source, have been sent back to India. Rest of it remains in Germany pending a decision on its safe disposal. One of the possible practices is to

82

use the items depending on their radiation levels in fencing or in bridges where the occupancy is less. Imported metal scrap After a thorough survey, the Atomic Energy Regulatory Board (AERB) concluded that the steel products in the recent incidents were made out of imported metal scrap which contained radioactive material. India imports more than 80 per cent of stainless steel metal scrap for recycling in the steel industry. Also most of the contaminated material was exported. The health consequences from these products were negligible, as the radiation levels were low. But the presence of even low radiation levels is not desirable. We do not have any estimate of the humongous economic losses including loss of business suffered by the industry. In some instances, the defaulters had to ship back the rejected material for safe disposal. In spite of regulatory control, radioactive sources get lost occasionally. These may be melted along with other metal scrap. The steel products include handle bars, manhole covers, metal straps, steel wires, lift buttons, metal strips used in leather bags etc. Sources licensed for use in India are unlikely to get into metal scrap, because of regulatory measures in place. However since there were a few instances of loss of control of sources, there is no room for complacency. Presently, we have no firm assurance that contaminated imported scrap will not enter the country. Several measures including the plan to install radiation monitors at shipping ports through which bulk of the imported scrap metals enter the country must be implemented swiftly. Precautions Every importer of metal scrap should obtain a certificate from the exporting country that the scrap is free from radioactivity. A multilayer radiation check system proposed by AERB should be followed to prevent the import and export of radioactive contaminated material. During this week, over 300 specialists from 62 countries including India are attending a five day International Conference on Control and Management of Inadvertent Radioactive Material in Scrap Metal, organized by the IAEA at Tarragona, Spain. The delegates asserted "that further steps are needed to protect people from radioactive material that can end up in junks and scrap yards," (IAEA release, Feb 23). All agencies must wake up before it is too late. Not long ago, explosion of live shells in imported metal scraps led to loss of life in India. Import of scrap laced with high levels of radioactivity is a possibility we must be concerned with.

K.S. PARTHASARATHY

FORMER SECRETARY, AERB"

Unquote.

7. The 2000s

"Cutting and other forms of self-mutilation may be hard to understand. People who self-harm, often have Borderline Personality Disorder (BPD). People with BPD may engage in self-injury because they get a sense of emotional relief from physical pain."
(According to German researchers.)

"I have a skin disorder that destroys the pigmentation of my skin, it's something that I cannot help, OK?"
Michael Jackson

"I have got this obsessive compulsive disorder where I have to have everything in a straight line, or everything has to be in pairs."
David Beckham

Wasted TV sets, PCs, mobile phones, and other kinds of electrical and electronic equipments from Norway (and other parts of the world) end up in landfills or incinerators, or are transported to Africa and Asia, where poor people are recycling and reusing metals and other components. *All these alternatives are causing special and general, local and global, manmade mutagenic pollution, and so are causing a general genetic catastrophe.*

Yes, I know the international waste directives and all that, but I also know that they do not work well.

When a modern house burns down enormous amounts of mutagenic and carcinogenic chemicals are produced and spread to the whole world, including the oceans. Some 1,000,000 mutagens end up in our food and our bodies. Norway has a world record when it comes to the number of burned down buildings.

In Norway's numerous road tunnels the concentrations of mutagenic and carcinogenic dust and gas are tragically high.

The towns Oslo and Bergen are two of the World's most polluted towns due to temperature inversions during the winters and Europe's oldest cars.

All the sea and air currents of the planet Earth meet outside the coast of Northern Norway. (Take a look at a Globe.). In these cold and productive waters mutagens from all over the world end up in the food chain. No wonder that Norway is the most gene-damaged nation on Earth.

The use of X-rays and scanning by Norwegian doctors and dentists are misused and out of control. The resulting gene damage is unknown. And no one cares. Quite the contrary; dentists take extra (unnecessary) X-ray pictures to earn more money.

The misuse and overuse of medicines in Norway is terrible and out of control. 80,000 Norwegians are totally addicted to medicines, and live short and bad lives, until their inner organs are destroyed. It is now increasingly clear that medicines, in general, are mutagenic. Also, the misuse and overuse of pesticides in Norway is terrible. Pesticides are, in general, mutagenic. But the politicians, the general population, and most scientists in Norway do not even know that tobacco smoke is mutagenic.

If you visit polluted work places in Norway you will find that the workers are badly protected. Example: Paving asphalt fumes are shown to be mutagenic and carcinogenic. But Norwegian asphalt workers do not use protection masks.

Electromagnetic smog from mobile phones, computers, etc. in schools and work places, is an additional genetic catastrophe. More and more research shows that electromagnetic smog is mutagenic and carcinogenic. Small children have their own mobile phones and PCs. The whole situation is a genetic experiment.

Extinction of a species is, of course, a genetic catastrophe. But this little book is not about extinction due to habitat loss, overfishing, and overhunting. This book is about genetic catastrophes caused by humanmade pollution, and mostly about genetic catastrophes caused by gene/genetic damage. It's basic ecology that all life is balancing on an edge of a knife. The escalating level of carbon dioxide in the atmosphere is making the world's oceans more acidic, killing corals and other marine organisms that secrete skeletal structures. *This global general genetic catastrophe, caused by manmade pollution, was not predicted by the scientists.*

In Norway hundreds of whole school classes (4 – 15 pupils) consist of gene damaged individuals (physically and/or mentally). These youths are unable to go through a normal curriculum. It's taboo to talk about these realities. The whole situation is kept secret. But also a large percentage of healthy students do have hidden or secret genetic conditions. When a student jump from a bridge, the newspapers are not allowed to report about it. Norwegian researchers, Statistics Norway, Norwegian politicians, and Norwegian schools, totally neglect gene/genetic damage to the population. (See section VI 9.)

In general, scientists do explain away the general genetic catastrophe of Chernobyl. But here comes a lengthy quote from *Annals of the New York Academy of Sciences* (2009):

Quote:

"Air particulate activity over all of the Northern Hemisphere reached its highest levels since the termination of nuclear weapons testing -- sometimes up to 1 million times higher than before the Chernobyl contamination. There were essential changes in the ionic, aerosol, and gas structure of the surface air in the heavily contaminated territories, as measured by electroconductivity and air radiolysis. Many years after the catastrophe aerosols from forest fires have dispersed hundreds of kilometers away. The Chernobyl radionuclides concentrate in sediments, water, plants, and animals, sometimes 100,000 times more than the local background level. The consequences of such a shock on aquatic ecosystems is largely unclear. Secondary contamination of freshwater ecosystems occurs as a result of Cs-137 and Sr-90 washout by the high waters of spring. The speed of vertical migration of different radionuclides in floodplains, lowland moors, peat bogs, etc., is about 2–4 cm/year. As a result of this vertical migration of radionuclides in soil, plants with deep root systems absorb them and carry the ones that are buried to the surface again. This transfer is one of the important mechanisms, observed in recent years, that leads to increased doses of internal irradiation among people in the contaminated territories.

Plants and mushrooms accumulate the Chernobyl radionuclides at a level that depends upon the soil, the climate, the particular biosphere, the season, spotty radioactive contamination, and the particular species and populations (subspecies, cultivars), etc. Each radionuclide has its own accumulation characteristics (e. g., levels of accumulation for Sr-90 are much higher than for Cs-137, and a thousand times less than that for Ce-144). Coefficients of accumulation and transition ratios vary so much in time and space that it is difficult, if not impossible, to predict the actual levels of Cs-137, Sr-90, Pu-238, Pu-239, Pu-240, and Am-241 at each place and time and for each individual plant or fungus. Chernobyl irradiation has caused structural anomalies and tumorlike changes in many plant species. Unique pathologic complexes are seen in the Chernobyl zone, such as a high percentage of anomalous pollen grains and spores. Chernobyl's irradiation has led to genetic disorders, sometimes continuing for many years, and it appears that it has awakened genes that have been silent over a long evolutionary time.

The radioactive shock when the Chernobyl reactor exploded in 1986 combined with chronic low-dose contamination has resulted in morphologic, physiologic, and genetic disorders in every animal species that has been studied—mammals, birds, amphibians, fish, and invertebrates. These populations exhibit a wide variety of morphological deformities not found in other populations. Despite reports of a "healthy" environment in proximity to Chernobyl for rare species of birds and mammals, the presence of such wildlife is likely the result of immigration and not from locally sustained populations. Twenty-three years after the catastrophe levels of incorporated radionuclides remain dangerously high for mammals, birds, amphibians, and fish in some areas of Europe. Mutation rates in animal populations in contaminated territories are significantly higher and there is transgenerational genomic instability in animal populations, manifested in adverse cellular and systemic effects. Long-term observations of both wild and experimental animal populations in the heavily contaminated areas show significant increases in morbidity and mortality that bear a striking resemblance to changes in the health of humans—increased occurrence of tumor and immunodeficiencies, decreased life expectancy, early aging, changes in blood and the circulatory system, malformations, and other factors that compromise health.

Of the few microorganisms that have been studied, all underwent rapid changes in the areas heavily contaminated by Chernobyl. Organisms such as tuberculosis bacilli; hepatitis, herpes, and tobacco mosaic viruses; cytomegalovirus; and soil micromycetes and bacteria were activated in various ways. The ultimate long-term consequences for the Chernobyl microbiologic biota may be worse than what we know today. Compared to humans and other mammals, the profound changes that take place among these small live organisms with rapid reproductive turnover do not bode well for the health and survival of other species."

Unquote.

A few parts per million (ppm) uranium occurs naturally in soil, rock, surface water, and ground water. So the soil of an ordinary farm contains several tons of uranium. Depleted uranium (DU) is Uranium-238. DU is used in armor-piercing munitions and protective tank plating. DU is a radioactive heavy metal, almost twice as dense as lead. A DU projectile begins to burn on impact, creating Uranium-238 dust. Of course, winds can transport the dust many miles. Inhalation of DU dust causes gene damage, chromosome damage, and chemical poisoning. (Remember: DU is a heavy metal.) An enormous number of hateful anti-US propaganda articles are found on the internet. I will make four comments on this:

1. The military use described above should stop
2. Science and scientists are the sinners here
3. Compared to the total mutagenic pollution by humans, DU pollution is marginal
4. Compared to the leftist sin of Chernobyl, DU use is almost nothing

In support of points 3. and 4. above, I bring the copy below.

Quote:

"SEATTLE POST-INTELLIGENCER, May 7, 2003
By DENIS D. GRAY, THE ASSOCIATED PRESS

Excerpt:

V Corps Command Sgt. Maj. Kenneth Preston said depleted uranium emits only extremely low levels of gamma radiation and low levels of alpha and beta particles that are easily blocked by skin and clothing.

He said passengers on a long airline flight are exposed to more radiation risk than soldiers hypothetically enclosed for one year in a tank surrounded by armor and shells with depleted uranium.

Sigmon said no special warnings have been issued to Iraqis about depleted uranium, but leaflets and other information are being circulated warning everyone, especially children, to stay away from all unexploded ordnance.

Preston said the vast ammunition and weapons caches of the former regime posed a far greater hazard than depleted uranium shells.

Preston, an expert on tank ammunition, said depleted uranium, a byproduct from the process of enriching uranium for use as fuel or in nuclear weapons, emits 40 percent less radiation than uranium occurring in its natural state. Depleted uranium is also 1.7 times heavier than lead, and thus does not stay suspended as dust particles for very long."

Unquote.

Children of the "liquidators" of Chernobyl (those drafted in to clear up the Chernobyl disaster) suffer *seven* times (raised 600 percent) the mutation rate of offspring whose parents were not exposed to radiation, according to research published in 2001 by the Royal Society. The researchers say that lower doses of radiation also produce mutations, suggesting that low level occupational or medical exposure to radiation could double the mutation rate in offspring. The *unexpectedly high* mutation rate means that *the large proportion of the world's population doing jobs where low-level radiation is present are exposing their unborn children to gene damage*, the researchers say. (Now we understand the leukaemia cluster at Sellafield. As we remember, science explained away that cluster.)

More than 2,000 former German soldiers now (2010) claim to have developed cancer after working on radar systems. The German military knew of the radiation risks, but did not warn soldiers to take necessary precautions until the early 1980s. Children of exposed soldiers are born with deformities linked to X-ray exposure.

Cancer incidence among some 27,000 diagnostic X-ray workers was compared to that of some 26,000 other medical specialists employed between 1950 and 1980 in China. X-ray workers had a 50 percent higher risk of developing cancer than the other specialists.

In a study published in the journal, *Acta Oncologica*, researchers have found an elevated risk of developing thyroid cancer in patients that have had repeated dental X-rays. The thyroid is extremely vulnerable to radiation carcinogenesis. It has been found that X-ray workers, dental assistants, and dentists have elevated risks of developing salivary tumors, thyroid cancer, and meningiomas due to dental X-rays.

Report from Israel in 2003: 5 patients (young military radar workers) had brain tumors that appeared within 10 years of initial occupational exposures to radar. Four of the patients were less than 30 years old when the diagnoses were initially made. Similarly, reports of short induction periods for brain cancer on the side of the head in which there has been prior use of cell phones also indicate increased risk of gene damage and cancer.

Now to Hiroshima and Nagasaki (2010: 65th anniversary of the Hiroshima / Nagasaki bombing.) Below I will give copies and quotes of several reports and articles on the problems of Hiroshima and Nagasaki.

The copy (quote) below is a copy from:

http://www.thehindu.com/sci-tech/article551620.ece?homepage=true

(NB: This article is published in The Hindu, August 5, 2010)

Quote:

"On August 6, 1945, the U.S. dropped an atom bomb on Hiroshima. Nagasaki was bombed three days later.

In the 1950 Japanese national census nearly 280,000 persons claimed that they were exposed to radiation. Initially, the Atomic Bomb Casualty Commission and from 1975, Radiation Effects Research Foundation (RERF), carried out several studies on the survivors.

Research programmes

The research programmes covered Life Span Study (LSS), Adult Health Study, study of the Children of Atomic-bomb Survivors (F{-1}) and the evaluation of the lifetime health experience of a specially exposed population, namely those *in utero* at the time of the bombings.

Other areas covered included immunology, radiation biology, molecular epidemiology, cytogenetics, statistics and A-bomb dosimetry.

RERF researchers and other scientists studied the interaction with radiation and smoking.

Radiation increased the risk of lung cancer among the survivors. Among 105,404 subjects of the LSS, researchers identified 1803 primary lung cancer cases for the period 1958-1999.

They used individual smoking history information and the latest radiation dose estimates to investigate the combined effects of radiation and smoking on lung cancer rates.

Lung cancer risks

Relative to never-smokers, lung cancer risks increased with the amount and duration of smoking and decreased with time since quitting smoking at any level of radiation exposure (*Radiation Research, 174, 2010*).

The excess risk increased rapidly with smoking intensity up to about 10 cigarettes per day, but additive or sub-additive for heavy smokers smoking a pack or more per day, with little indication of any radiation-associated excess risk.

The authors concluded that the joint effect of smoking and radiation on lung cancer in the LSS is dependent on smoking intensity and is best described by the generalized interaction model rather than a simple additive or multiplicative models.

Fatty liver predicts ischemic heart disease or heart disease due to reduced blood supply to the heart.

Fatty liver predictors

The researchers at RERF observed the incidence and predictors of fatty liver by examining 1635 survivors of Nagasaki A-bomb every two years through 2007 (mean follow up for 11.6 y) by abdominal ultrasonography.

The subjects were without fatty liver at base line (November 1990 through 1992). The researchers diagnosed 323 new fatty liver cases.

The average incidence was 19.9 cases in 1000 person years peaking in the sixth decade of life (*Hypertension Research* April 2010).

After controlling for age, sex, and smoking and drinking habits, obesity, hypertriglyceridemia (large levels of tryglicerides) and hypertension were predictive of fatty liver.

All variables included

When all variables are included, obesity, hypertriglyceridemia and hypertension remained predictive.

Scientists have not observed genetic effects in the children of A-bomb survivors.

To evaluate the genetic effects of A-bomb radiation, RERF researchers examined mutations at specific loci in the chromosomes of exposed families (father-mother-offspring, mostly uni-parental exposures) and control families. The mutation rates observed were not statistically significant.

That radiation exposure causes thyroid cancer is an established fact.

But we do not know the radiation effects on papillary micro carcinoma (PMC) of the thyroid, a common sub-clinical thyroid malignancy.

RERF researchers identified PMCs in a subset of 7659 subjects after reviewing their pathology and evaluating the histological features of the tumors.

Papillary thyroid cancer

From 1958 to 1995, they detected 458 PMCs among 313 study subjects; most of them exhibited pathologic features of papillary thyroid cancers.

A significant radiation-dose response was found for the prevalence of PMCs with the excess risk observed primarily among women.

Exposure to low-to-moderate doses of ionizing radiation appears to increase the risk of thyroid PMCs, even when exposure occurs during adulthood (*Cancer, 2010*).

Raja Ramanna Fellow with the Department of Atomic Energy.

(ksparth@yahoo.co.uk)."

Unquote.

The copy (quote) below is a copy from:

http://www.counterpunch.org/johnston08112010.html

(NB: This (part of an)article *and* The Hindu Times article (see above) are probably partly propaganda articles, but with opposite purposes. The reader should read both articles carefully to try to understand the reality of radioactive pollution.)

Quote:

"An article from The Hindu Times on the 65th anniversary of the Hiroshima bombing illustrates this "trust us, science has shown nuclear fears are overblown" mantra. LINK here ("No genetic effects in children of A-bomb survivors" by K. S. Parthasarathy, a former Secretary of India's Atomic Energy Regulatory Board). The article begins by mentioning that in 1950 some 280,000 Japanese "claimed they were exposed to radiation." It then briefly outlines a few studies conducted by the Radiation Effects Research Foundation (RERF) and its predecessor the Atomic Bomb Casualty Commission (ABBC) on the lifetime health experience of exposed survivors and their children, with specific attention to lung cancer, hypertension and heart disease, and thyroid cancer. For lung cancer and heart disease, incident rates were largely attributed to lifestyle: "Scientific study has determined that in these specific disease processes there is little indication of any radiation-associated excess risk." The relationship between radiation exposure and thyroid disease (a treatable condition), the author acknowledges: "that exposure causes thyroid cancer is an established fact." The article concludes with the assertion that "scientists have not observed genetic effects in the children of A-bomb survivors" a statement based on the ABCC's evaluation of "mutations at specific loci in the chromosomes" of exposed families and control families, and findings that the mutation rates observed were not statistically significant...".

The subtext asserted here? A casual reader might conclude nuclear war, while deadly, does not present a lingering nor inter-generational effect. The informed reader might do further research to find that the author of this article, K. S. Parthasarathy, may have a vested interest in such a framing, as a former Secretary of India's Atomic Energy Regulatory Board. And, an obsessed reader, such as myself, might wonder about the source material.

The supporting study is J.V. Neel's oft-cited assessment of Japanese pregnancy and outcomes data on congenital defects for the years 1948-54, collected by the ABCC with a protocol designed by Neel. The initial assessment, published by Neel and his associate WJ Schull, found a statistically relevant change in the sex ratio. I first came across this study while going through the National Academy of Science archives on the Atomic Bomb Casualty Commission. Those files included copies of Neel's research protocol, the study findings after five years of data collection, and the National Research Council critique of methods and conclusions, including the noted concern that a flawed control group might have been constituted. In this initial assessment of abnormal outcomes Neel assumed most birth defects to be the consequence of inbreeding and thus he added a discount factor to the protocol (rejecting data from parents who were first, second, or third cousins). Lacking pre-exposure data to compare outcome rates, the study relied upon a control population including people who resided outside of the blast zone (assuming fallout in the broader region is not relevant) and including residents of Kure, a town (visible from Hiroshima ground zero). Years later, in the RERF newsletter Update, Neel recalled that "a drastic drop in the live birth rate in 1954 led to the discontinuation of physical exams, though data on sex ratio and survival of newborns was still collected."

The conclusions reached in this influential, yet flawed study, reflect the context of the times, state of knowledge, and cultural biases about populations and consanguinity (the distance between blood relatives) that permeated US science in the 1940s and 50s-era. The policy implications of this study have had profound impact in explaining away experience and liability with regards to the reproductive and inter-generational effects of radiation exposure. For example, this study of Neel's was cited as the scientific rationale for denying further study and treatment in the Marshall Islands when complaints of birth defects and other reproductive problems were filed decades ago."

Unquote.

Certainly, the problems of Hiroshima and Nagasaki are extremely difficult. So let us therefore study more reports on the problems.

The copy (quote) below is a copy from:

http://www.ncbi.nlm.nih.gov/pubmed/20095853

(NB: This article is a 2010 article.)

Quote:

"Radiat Res. 2010 Feb;173(2):205-13.

No evidence of increased mutation rates at microsatellite loci in offspring of A-bomb survivors.

Kodaira M, Ryo H, Kamada N, Furukawa K, Takahashi N, Nakajima H, Nomura, T, Nakamura N.

Departments of Genetics and, Radiation Effects Research Foundation, Hiroshima, Japan. kodaira@rerf.or.jp

Abstract

To evaluate the genetic effects of A-bomb radiation, we examined mutations at 40 microsatellite loci in exposed families (father-mother-offspring, mostly uni-parental exposures), which consisted of 66 offspring having a mean paternal dose of 1.87 Gy and a mean maternal dose of 1.27 Gy. The control families consisted of 63 offspring whose parents either were exposed to low doses of radiation (< 0.01 Gy) or were not in the cities of Hiroshima or Nagasaki at the time of the bombs. We found seven mutations in the exposed alleles (7/2,789; mutation rate 0.25 x 10(-2)/locus/generation) and 26 in the unexposed alleles (26/7,465; 0.35 x 10(-2)/locus/generation), which does not indicate an effect from parental exposure to radiation. Although we could not assign the parental origins of four mutations, the conclusion may hold since even if we assume that these four mutations had occurred in the exposed alleles, the estimated mean mutation rate would be 0.39 x 10(-2) in the exposed group [(7 + 4)/2,789)], which is slightly higher than 0.35 x 10(-2) in the control group, but the difference is not statistically significant.

PMID: 20095853 [PubMed - indexed for MEDLINE]"

Unquote.

My (NKO's) comment: The results reported here clearly show that the leftists are misusing Fallujah (see next copied report just below), Hiroshima, and Nagasaki for anti-US propaganda purposes.

The copy (quote) below is from the internet.

Quote:

"Int. J. Environ. Res. Public Health2010, 7, 2828-2837; doi:10.3390/ijerph7072828

International Journal of
Environmental Research and
Public Health
ISSN 1660-4601
www.mdpi.com/journal/ijerph

Article

Cancer, Infant Mortality and Birth Sex-Ratio in Fallujah, Iraq

2005–2009

Chris Busby1 ,*, Malak Hamdan2 and Entesar Ariabi3

1Department of Molecular Biosciences, University of Ulster, Cromore Rd, Coleraine,

BT52 1SA, UK

2100 Tanfield Avenue, Neasden, London, NW2 7RT, UK;
E-Mail: malakhamdan@hotmail.com

382 Goldsmith Road, London, N11 3JN, UK; E-Mail: intisar_
alobady@yahoo.com

* Author to whom correspondence should be addressed; E-Mail:
christo@greenaudit.org;
Tel.: +44-1970-630215; Fax: +44-1970-630215.

Received: 7 June 2010; in revised form: 23 June 2010 / Accepted:
30 June 2010 /
Published: 6 July 2010

Abstract: There have been anecdotal reports of increases in birth defects and cancer in Fallujah, Iraq blamed on the use of novel weapons (possibly including depleted uranium) in heavy fighting which occurred in that town between US led forces and local elements in 2004. In Jan/Feb 2010 the authors organised a team of researchers who visited 711 houses in Fallujah, Iraq and obtained responses to a questionnaire in Arabic on cancer, birth defects and infant mortality. The total population in the resulting sample was 4,843 persons with and overall response rate was better than 60%. Relative Risks for cancer were age-standardised and compared to rates in the Middle East Cancer Registry (MECC, Garbiah Egypt) for 1999 and rates in Jordan 1996–2001. Between Jan 2005 and the survey end date there were 62 cases of cancer malignancy reported (RR = 4.22; CI: 2.8, 6.6; p < 0.00000001) including 16 cases of childhood cancer 0-14 (RR = 12.6; CI: 4.9, 32; p < 0.00000001). Highest risks were found in all-leukaemia in the age groups 0-34 (20 cases RR = 38.5; CI: 19.2, 77; p < 0.00000001), all lymphoma 0–34 (8 cases, RR = 9.24;CI: 4.12, 20.8; p < 0.00000001), female breast cancer 0–44 (12 cases RR = 9.7;CI: 3.6, 25.6; p < 0.00000001) and brain tumours all ages

(4 cases, RR = 7.4;CI: 2.4, 23.1; P < 0.004). Infant mortality was based on the mean birth rate over the 4 year period 2006–2009 with 1/6th added for cases reported in January and February 2010. There were 34 deaths in the age group 0–1 in this period giving a rate of 80 deaths per 1,000 births. This may be compared with a rate of 19.8 in Egypt (RR = 4.2 p < 0.00001) 17 in Jordan in 2008 and 9.7 in Kuwait in 2008. The mean birth sex-ratio in the recent 5-year cohort was anomalous. Normally the sex ratio in human populations is a constant with 1,050 boys born to 1,000 girls. This is disturbed if there is a genetic damage stress. The ratio of boys to 1,000 girls in the 0–4, 5–9, 10–14 and 15–19 age cohorts in the Fallujah sample were 860, 1,182, 1,108 and 1,010 respectively suggesting genetic damage to the 0–4 group (p < 0.01). Whilst the results seem to qualitatively support the existence of serious mutation-related health effects in Fallujah, owing to the structural problems associated with surveys of this kind, care should be exercised in interpreting the findings quantitatively.

Keywords: Fallujah; Iraq; cancer; leukemia; depleted uranium; gulf war."

Unquote.

My (NKO's) comment: This article is widely misused on the internet for anti-US propaganda purposes. Moreover this report is perhaps a bit propagandistic itself. Nevertheless, we give more Fallujah quotes:

The four copies (quotes) below are copied from:

http://www.indymedia.org.nz/article/78827/fallujah-%E2%80%93-poisoned-city.

(NB: Most of the total article (see link above) was simply stupid anti-US propaganda. The four quotes below are therefore carefully selected from the article. So I do think the quotes below are really informative, interesting, and important.)

Quote:

"The DU oxide dust produced when DU munitions burn has no natural or historical analogue. This toxic and radioactive dust is composed of two oxides: one insoluble, the other sparingly soluble. The distribution of particle sizes includes sub-micron particles that are readily inhaled into and retained by the lungs. From the lungs uranium compounds are deposited in the lymph nodes, bones, brain and testes. Hard targets hit by DU penetrators are surrounded by this dust and surveys suggest that it can travel many kilometres when re-suspended, as is likely in arid climates. The dust can then be inhaled or ingested by civilians and the military alike. [...]

We are seeing a very significant increase in central nervous system anomalies," said Falluja general hospital's director and senior specialist, Dr Ayman Qais. "Before 2003 [the start of the war] I was seeing sporadic numbers of deformities in babies. Now the frequency of deformities has increased dramatically. [...]

The effects of DU are just beginning as it has a half-life of 4.5 billion years. [...]

As Dahr Jamail reported in 08, "depleted uranium (DU) munitions, which contain low-level radioactive waste, were used heavily in Fallujah. The Pentagon admits to having used 1,200 tonnes of DU in Iraq thus far.","

Unquote.

My (NKO's) comment: I do think that the population of Fallujah must be intensely studied for the next 50 – 100 years (or more), just as Japan and the US are doing in Hiroshima and Nagasaki. Of course, the GGC is at work in Fallujah as well. That fact makes the Fallujah problems extremely difficult.

Finally, let's study this report:

The copy (quote) below is from Health Physics Society at:

http://www.solarstorms.org/Hiroshima.html

Quote:

"Answer to Question #340 Submitted to "Ask the Experts" [Answer posted on July 28, 2000.]

Category: Radiation Effects—Hiroshima and Nagasaki

The following question was answered by an expert in the appropriate field:

A:
Many Japanese died immediately from the blast forces, heat, and fires resulting from the atomic explosions at Hiroshima and Nagasaki, and a large number died within weeks or months from radiation effects. (1) What information is available about the health and life-span of those who survived? (2) For example, how many were still alive after 50 years and how does this compare with Japan in general?

(3) What about genetic effects in the children born of the survivors? (4) Is there reasonable evidence from the work with survivors to support the possibility of low radiation being less harmful than expected or even beneficial? Thanks for your expert service.

B:

An excellent reference for all manner of questions regarding the A-bomb survivors is the book by William J Schull, *Effects of Atomic Radiation: A Half-Century of Studies from Hiroshima and Nagasaki*, Wiley-Liss, Inc., 605 Third Avenue, New York, NY 10158-0012 (1995) ; ISBN # 0-471-12524-5. This is a scholarly book and yet is written at a level that the intelligent layperson can understand. It has a wealth of historical as well as scientific information about the studies, spanning the entire time period since the bombs were dropped. Following are answers to the specific questions asked:

1. Much information is available; for example, in the book cited above.

2. In 1995, 50 years after the atomic bombings, approximately 50 percent of the survivors were still alive. The exact number is difficult to state, but it could exceed 100,000. (For example, 284,000 survivors were identified in the 1950 census; this would indicate that there were about 142,000 remaining survivors in 1995.)

3. No genetic effects have been detected in a large sample (nearly 80,000) of offspring. By this, we mean that there is no detectable radiation-related increase in congenital abnormalities, mortality (including childhood cancers), chromosome aberrations, or mutations in biochemically identifiable genes.

4. Unfortunately, the epidemiologic studies on the survivors who received low doses of radiation (in the range of 0.01 Sv to 0.2 Sv) are equivocal regarding good measures of the risk of long-term health effects. This is because, even though the statistical sample available in the survivor studies is very large (nearly 100,000 subjects in the Life Span Study), it can be shown that many, many more subjects would be needed to draw reasonable statistically valid inferences from the data. Thus the data at low doses have large error bars and can be fit to mathematical models that show a threshold, no threshold, reduced effect, and in some cases even a beneficial (protective) effect, depending on the model one picks. There is no model that seems to be more valid than the others. Therefore, the consensus of the community of scientists interested in the A-bomb, as well as other, radiation studies seems to be that epidemiologic studies do not have the statistical power to give us answers to the low-dose questions. This issue is thoroughly discussed in the book by William J. Schull.

John D. Zimbrick, Ph.D. School of Health Sciences Purdue Univerity

Treatment of victims of nuclear bombs depends on how close they were to the hypocenter and numerous other factors. Those in close proximity were killed acutely by blast and heat, and no treatment was possible. Those a little farther away, who survived these effects, suffered from acute radiation syndromes and became quite ill within hours to days. The duration of the latent period is inversely proportional to radiation dose, that is, proximity to the hypocenter. In Japan most of these received little or no treatment. The problem was the massive number of casualties in this class and the lack of knowledge of the medical personnel. In cases of nuclear accidents (for example, Chernobyl) the subjects who got doses greater than 10-12 Gy received supportive care only (nutrition, fluids, narcotics for pain, etc.) because their radiation injuries were universally fatal. Those in the 8-10 Gy range could benefit from marrow transplants. Those below 8 Gy would probably survive with supportive care. Exposed individuals who survived the acute effects, however, were later found to suffer increased incidence of cancer of essentially all organs. The cancers occurred years to decades later. Excess cancers are still being detected in this population, now more than 50 years after the bombing. Excess cancer means that these individuals are more likely to get cancer than other Japanese. The cancers they get are in no way different from spontaneous cancer in other Japanese. Animal studies have detected genetic effects from these sublethal doses: mutations that occur in offspring, perhaps several generations later. No such effects have been detected in offspring of Japanese survivors. However, most mutations are recessive and require several generations to detect. The second generation of offspring of the Japanese is just now appearing.

S. Julian Gibbs, DDS, PhD"

Unquote.

My (NKO's) comment: *Point 3. above is an absolutely wrong answer. Of course, there is a lot of damage to the DNA of children of A-bomb survivors.*

Below I bring a nightmare breaking news quoted from NaturalNews.com.

Quote:

"Nanoparticles may be able to damage the DNA of cells without ever coming into contact with it, according to a study conducted by researchers from the Bristol Implant Research Center and published in the journal *Nature Nanotechnology*.

Nanoparticles are particles so small that they have fundamentally different physical and chemical properties than the same substances do at more familiar scales. Industry is increasingly adopting nanotechnology for a variety of applications, from consumer products c to medicine, but the technology remains unregulated.

Researchers created particles of chromium and cobalt that were either four millionths (micro scale) or 30 billionths (nano scale) of a meter across, then placed them on a thin, artificial membrane composed of human cells. On the other side of the membrane, researchers placed human fibroblast cells, which are important components of connective tissue.

They found that although no particles crossed the cellular membrane, fibroblast cells placed across from the metal particles suffered DNA damage in 10 times as many locations and cells placed next to a membrane with nothing on the other side.

Researchers are unsure how the particles damaged the cells without crossing the membrane, but they believe they may cause changes in the membrane cells, which in turn signal the fibroblast cells and cause DNA damage.

"We used a variety of chemicals to block ... cell-to-cell signaling and found that in the presence of these blockers, the damage we were seeing was completely prevented," lead author Gevdeep Bhabra said.

The experiment was conducted with cobalt and chromium because both of those metals are currently used in medical implants. The researchers noted, however, that it would be unlikely for wear and tear to produce enough nano- or micro-sized particles to reach the concentrations used in the study. The implications of the study center more around the risks of actual nanotechnology.

Nanoparticles are already used in the manufacture of sunscreens, cosmetics, sporting goods and other consumer products. Researchers are also investigating their use as drug-delivery mechanisms.

Sources for this story include: news.bbc.co.uk."

Unquote.

What is the *total* result of the mutagenic Hell of the Modern world? Well, the totality of information found in this book tries to answer this very question. Here, at the end of this section of the book, I only want to focus of one particular segment of the human body, namely the external genitals in *children*. The findings shown below are beyond imagination and overwhelmingly tragic.

External genital abnormalities (examples from Serbia and Egypt):

"A total of 1229 elementary school boys were examined. The incidence of external genital abnormalities was 27.8%. Certain anomalies were already surgically treated in 7.8% of boys. Phimosis

was found in 66 patients (5,5%), which represents 26.6% of all abnormalities."
(Serbian report, 2004, published at PubMed.gov.)

"External genital anomalies are among the most common congenital anomalies. Proper early diagnosis and management of genital abnormalities are of great importance to minimize medical, psychological and social complications. AIM: To detect the incidence of external genital anomalies and disorders of sex development (DSD) in Great Cairo and Qalyubiyah governorates. SUBJECTS AND METHODS: 20,000 newborns and infants up to the age of 6 months coming for compulsory vaccination at primary health care units and centers in Great Cairo and Qalyubiyah governorates were examined in the years 2006-2007 for suspected genital anomalies. RESULTS: There were 187 (93.5/10,000) cases with external genital anomalies among the screened 20,000 participants. Various abnormalities in the form of 46,XY DSD, undescended testis, hydrocele, hypospadias, micropenis, synechia of the labia and other genital anomalies were diagnosed and classified after thorough clinical examination, and hormonal, radiological, and laparoscopic investigations."
(Egyptian report, published at BioMedSearch.com.)

Olyslager and Conway presented a paper on transsexuals at the WPATH 20th International Symposium in 2007. They suggest the prevalence of transsexuals might be as high as 1:500 births overall. But if we look at the flow of thousands of *beautiful* shemales (women with perfect female body *and* large perfect penises), for example, in the porn industry, I do think the prevalence may be even higher than 1 : 500.

8. My 50 types of gene damage

The author has the following known types of gene damage. Of course, the list is extremely incomplete because only known damage is listed here. Perhaps, such a *personal* list of gene damage has never before been published by a researcher. *I do think the list below represents the founding of a new science or a new field of research.*

1. I have several types of lipomas all over the body.

2. I have fibromas all over the body. They are situated deeper than the liopmas. Some of them are causing pain when touched.

3. Hypodermis is generally sore, and I don't like to be scratched on my skin.

4. Severe dandruff (genetic condition).

5. My skin don't like sunshine and heat radiation.

6. Excessive sweating (too many sweat glands or they are abnormally large).

7. Bad sweat smell (inherited from my mother).

8. Numbness in left side of left thigh.

9. White fingers.

10. White toes.

11. The acoustic meatus of my left ear is much contracted.

12. The elbows are different (the olecranons are different).

13. Trouble with my Achilles' tendons.

14. Left kidney has two ureters.

15. Urine jet is out of control (probably anomalous anatomy).

16. Chewing food with partially open mouth (inherited from my father).

17. Eating soup with a spoon: A part of the soap flows down the chin. (Anomalous anatomy of the tongue?).

18. Constipation problems (inherited from my father).

19. Bad breath (from anatomic stomach problems?).

20. Unexplained overactive urinary bladder.

21. Unexplained prostate enlargement.

22. Damaged automatic voice volume control: Therefore excessively loud speaking voice.

23. Bad defecation smell (outside the normal range; inherited from my mother).

24. Extreme drinking water requirement. (See genetic condition 6.)

25. Bad temperature regulation in my body.

26. I've been toppling my milk glass (and other objects.) on the table since I was a child. (Probably a genetic condition.)

27. Confusing left and right. (I must think!)

28. Confusing (exchanging) North and South.

    (My biological compass is 180 degrees wrong.)

29. Confusing (exchanging) East and West.

30. Confusing (exchanging) Tuesday and Friday.

31. It's difficult to know if the road is horizontal, slowly descending, or slowly ascending.

32. Bad in mental arithmetic. (But I'm good in advanced mathematics.)

33. Extremely unmusical. (Inherited from my mother.)

34. Do not remember names and numbers. (But I do remember faces extremely well.)

35. Bad, varying, and chaotic handwriting.

36. Cannot relax properly.

37. Panic too easily. (Inherited from my mother.) Only a little panic makes me incapable of making proper judgements.

38. Something strange happens to my brain a few times a year and lasts for approximately 10 – 30 seconds.

39. Cannot stay focused on boring things. Example: It's almost impossible to catch the weather report.

40. Extremely bad chess player, scrabble player, etc. (Even an unexperienced child beats me.)

41. Extremely carsick. (Outside the normal range.)

42. The left buttock is allergic to the right buttock.

43. Pollen allergy (weak).

44. Food intolerance to citrus fruits.

45. Food intolerance to meat.

46. Food intolerance to egg.

47. Food intolerance to milk. (Inherited from my maternal grandfather.)

48. Food intolerance to almond.

49. Wasp allergy. (Outside the normal range.)

50. Extreme flatulence caused by food fibers. (Outside the normal range.)

9. Seven pairs of historically interesting letters

1th pair of letters (question and answer)

Question:

Nils K. Oeijord
Styrmannsv 62
9014 Tromso   2002-04-21
Norway

Forskningsavdelingen
Statistisk sentralbyrå
Postboks 8131 Dep.
0033 Oslo

Genetic damage of genetic disorders, facts and information

Could you kindly "print out" the following information on each of the several thousand genetic disorders of the human population of Norway, and send it to me:

Year the genetic disorder/genetic disease/genetic damage was first detected (in Norway)

Number of new cases per year

Total number of cases

Also, could you kindly send me a copy of the prognoses of the Statistisk sentralbyrå for each of the genetic disorders for the next 100 years.

Please answer in English because your answer is extremely important and will be used in my next book on the genetic catastrophe. If I've to pay for this service, it's OK.

Most sincerely,
Nils K. Oeijord

Answer:

Statistisk sentralbyrå
Statistics Norway

Nils K. Oeijord
Styrmannsv. 62
9014 Tromso

Oslo, 21 June 2002

Your ref.: , Our ref.: 02/954-1
Executive officer: Anne Mundal
Division for Health Statistics

Thank you for your request about information on genetic disorders. Unfortunately we don't have the information that you ask for, but we recommend that you contact Norwegian Institute of Public Health, P.O. Box 4404 Nydalen, N-0403 OSLO, telephone 22 04 22 00.

Best regards,
Anne Mundal

2nd pair of letters (question and answer)

Question:

Nils K. Oeijord
Styrmannsveien 62
9014 Tromso   2002-08-05
Norway

Telephone/fax +47 776 71 215
Mobile phone +47 99 10 13 86
E-mail n-oeij@frisurf.no

Norwegian Institute of Public Health
P.O. Box 4404 Nydalen, N-0403 OSLO
Norway

Genetic damage of genetic disorders, facts and information

Could you kindly "print out" the following information on each of the several thousand genetic disorders of the human population of Norway, and send it to me:

Year the genetic disorder/genetic disease/genetic damage was first detected (in Norway)

Number of new cases per year

Total number of cases

Also, could you kindly send me a copy of the prognoses for each of the genetic disorders for the next 100 years.

Please answer in English because your answer is extremely important and will be used in my next book on the genetic catastrophe. If I've to pay for this service, it's OK.

Very sincerely,
Nils K. Oeijord

Answer:

Folkehelseinstituttet

Nils K. Oeijord
Styrmannsv 62
9014 Tromso

Dear Nils K. Oeijord

We have received your letter sent August 5 2002. Unfortunately, there is no registration centrally of cases of genetic diseases, so we have no way of answering your question. As far as we know there is no indication that the frequency of genetic diseases is increasing in Norway.

Sincerely yours,
Per Magnus, MD
Head of Department

3rd pair of letters (question and answer)

Question:

Nils K. Oeijord
Styrmannsveien 62
9014 Tromso   2002-08-10
Norway

Telephone/fax +47 776 71 215
Mobile phone +47 99 10 13 86
E-mail n-oeij@frisurf.no

Norwegian Institute of Public Health
P.O. Box 4404 Nydalen, N-0403 OSLO
Norway

Non-central registration of genetic diseases/genetic disorders/genetic damage in Norway.

Thank you very much for your answer of August 6 2002 to my letter sent August 5 2002. Where can I find non-central registration of genetic diseases/genetic disorders/genetic damage in Norway?

Very sincerely,
Nils K. Oeijord

Answer:

Folkehelseinstituttet

Nils K. Oeijord
Styrmannsv 62
9014 Tromso

September 11, 2002

Dear Nils K. Oeijord

Thank you for your letter sent August 10 2002. Patients with genetic diseases will some times attend the Departments of Medical Genetics at our four Universities. Additionally, they will seek help from general practitioners. However, I doubt whether they can be of much help in giving numbers that will help in judging the occurrence of genetic disorders in Norway.

Sincerely yours,
Per Magnus, MD
Head of Department

4th pair of letters (question and answer)

Question:

Nils K. Oeijord
Styrmannsveien 62
9014 Tromso   2002-09-16
Norway

Telephone/fax +47 776 71 215
Mobile phone +47 99 10 13 86
E-mail n-oeij@frisurf.no

Department of Medical Genetics
University of Oslo
P. O. Box 1130 Blindern
0318 Oslo

Genetic damage of genetic disorders, facts and information

Could you kindly "print out" the following information on each of the several thousand genetic disorders of the human population of Norway, and send it to me:

Year the genetic disorder/genetic disease/genetic damage was first detected (in Norway)

Number of new cases per year

Total number of cases

Also, could you kindly send me a copy of the prognoses for each of the genetic disorders for the next 100 years.

Please answer in English because your answer is extremely Important and will be used in my next book on the genetic catastrophe. If I've to pay for this service, it's OK.

Very sincerely,
Nils K. Oeijord

Answer (E-mail):

[September 19, 2002]

Re your letter of 2002-09-16,

Unfortunately the information you ask for is not available in Norway (or in any other country, I believe).

Sincerely yours,

Carl Birger van der Hagen
Prosektor, overlege

Institutt for medisinsk genetikk
Universitetet i Oslo
Boks 1036
0315 Oslo
tlf +47 22 11 98 81

5th pair of letters (question and answer)

Question:

Nils K. Oeijord
Styrmannsveien 62
9014 Tromso   2002-09-16
Norway

Telephone/fax +47 776 71 215
Mobile phone +47 99 10 13 86
E-mail n-oeij@frisurf.no

Department of Medical Genetics
University of Bergen
Jonas Liesvei 65
5021 Bergen

Genetic damage of genetic disorders, facts and information

Could you kindly "print out" the following information on each of the several thousand genetic disorders of the human population of Norway, and send it to me:

Year the genetic disorder/genetic disease/genetic damage was first detected (in Norway)

Number of new cases per year

Total number of cases

Also, could you kindly send me a copy of the prognoses for each of the genetic disorders for the next 100 years.

Please answer in English because your answer is extremely important and will be used in my next book on the genetic catastrophe. If I've to pay for this service, it's OK.

Very sincerely,
Nils K. Oeijord

Answer:

Haukeland University Hospital
Center for medical genetics and
molecular medicine

Mr. Nils K. Oeijord
Styrmannsveien 62
9014 Tromso   Bergen, 17. September 2002

Genetic damage of genetic disorders, facts and information

In response to your inquiry received here today, I'm sorry to inform you that no such data exist. Moreover, there are no ways to generate such data in a reliable way, regardless of resources put into such an undertaking. We lack data even for relatively "common" genetic disorders, as there is no mandatory registration of (genetic) disease in Norway (apart from cancer).

Our Surgeon General (Helsedirektoeren) recommended in the 1950'ies that some registration of selected disorders (considered inherited at that time) should take place at the Institute of Medical Genetics, University of Oslo. Thus, there may still be some registrations of inherited disease in the archives of this institute. This registration was never properly funded, was not compulsory and was completely abandoned many years ago. The head of the Institute of Medical Genetics at the time this registration was initiated was professor Jan Mohr (appointed later on to the Chair in medical genetics, University Of Copenhagen). The head of the registry ("Arvelighetsregistret") throughout most of the registration period was professor Kaare Berg, recently retired.

Sincerely,
Helge Boman, M.D.
Professor of Medicine (Medical Genetics)
University of Bergen

6th pair of letters (question and answer)

Question:

Nils K. Oeijord
Styrmannsveien 62
9014 Tromso   2002-09-16
Norway

Telephone/fax +47 776 71 215
Mobile phone +47 99 10 13 86
E-mail n-oeij@frisurf.no

Department of Medical Genetics
University of Tromso
9037 Tromso

Genetic damage of genetic disorders, facts and information

Could you kindly "print out" the following information on each of the several thousand genetic disorders of the human population of Norway, and send it to me:

Year the genetic disorder/genetic disease/genetic damage was first detected (in Norway)

Number of new cases per year

Total number of cases

Also, could you kindly send me a copy of the prognoses for each of the genetic disorders for the next 100 years.

Please answer in English because your answer is extremely important and will be used in my next book on the genetic catastrophe. If I've to pay for this service, it's OK.

Very sincerely,
Nils K. Oeijord

Answer:

They did answer in Norwegian. Moreover, the answer was worthless. First, they even used the telephone to avoid writing a letter.

7th pair of letters (question and answer)

Question:

Nils K. Oeijord
Styrmannsveien 62
9014 Tromso   2002-09-16
Norway

Telephone/fax +47 776 71 215
Mobile phone +47 99 10 13 86
E-mail n-oeij@frisurf.no

Department of Medical Genetics
Norwegian University of Science and Technology (NTNU)
7491 Trondheim

Genetic damage of genetic disorders, facts and information

Could you kindly "print out" the following information on each of the several thousand genetic disorders of the human population of Norway, and send it to me:

Year the genetic disorder/genetic disease/genetic damage was first detected (in Norway)

Number of new cases per year

Total number of cases

Also, could you kindly send me a copy of the prognoses for each of the genetic disorders for the next 100 years.

Please answer in English because your answer is extremely important and will be used in my next book on the genetic catastrophe. If I've to pay for this service, it's OK.

Very sincerely,
Nils K. Oeijord

Answer:

They did not answer.

My comments to the answers above: According to EURORDIS: "A disease or disorder is defined as rare in Europe when it affects less than 1 in 2,000. One rare disease may affect only a handful of patients in the EU, and another touch as many as 245,000. There are between 6,000 and 8,000 rare diseases. On the whole, rare diseases may affect 30 million European Union citizens. 80% of rare diseases are of genetic origin, and are often chronic and life-threatening." According to the letters above, obviously, *Norwegian researchers were boycotting the whole field of genetic diseases.* But every Norwegian parent dreads learning that his/her child is ill: heart disease, digestive system disease, blood disease, skin disease, and so on. By the way, here comes a nightmare breaking news report:

Lancet, Volume 376, Issue 9750, Pages 1401 - 1408, 23 October 2010.
Published Online: 30 September 2010:

Rare chromosomal deletions and duplications in ADHD!!

10. The autism explosion is nothing special

"People with autism can be a little autistic or very autistic. Thus, it is possible to be bright, verbal, and autistic as well as mentally retarded, non-verbal and autistic. A disorder that includes such a broad range of symptoms is often called a spectrum disorder; hence the term "autism spectrum disorder." The most significant shared symptom is difficulty with social communication (eye contact, conversation, taking another's perspective, etc.)."

Lisa Jo Rudy (About.com, reviewed by the Medical Review Board)

"Asperger Syndrome (AS) is considered to be a part of the autism spectrum. The only significant difference between AS and High Functioning Autism is that people with AS usually develop speech right on time while people with autism usually have speech delays. People with AS are generally very bright and verbal, but have significant social deficits (which is why AS has earned the nickname "Geek Syndrome")."

Lisa Jo Rudy (About.com, reviewed by the Medical Review Board)

The *total* natural mutation rate in humans is ca. 70 new mutations (in coding and non-coding genes) per newborn (per fertilized egg). Today, in the GGC environment, the unnatural mutation rate is perhaps several hundred, or several thousand new mutations per newborn per genome. And the genes of the repair enzymes, the genes of the apoptosis system, etc., are damaged, causing a genetic domino effect. The manmade mutagens are particularly adapt at causing certain spectrums of diseases. One of these spectrums is autism. Therefore the autism explosion is not a mystery. Autism is no more mysterious than the cancer spectrum, the diabetes spectrum, the Alzheimer spectrum, etc. All genetic diseases are increasing, and new genetic diseases are popping up all the time, just as expected in the manmade super-mutagenic environment of the GGC. Diabetes explosion in Norway: New research from the Norwegian Diabetes Association measures explosive growth (some 400 percent) over the past 50 years. Norwegian Diabetes Association is worried about the explosive growth and expects a further doubling over the next 20 years if nothing is done. One more explosion: Every year some 500 children (increasing number) with congenital heart defects are born in Norway, and Norway has only some 60,000 births per year. Note that the autism explosion is not stronger than, for example, the diabetes explosion or the heart defect explosion in newborns. (See below.)

"The backdrop here is that autism rates are skyrocketing in American children. My InvestigateWest colleague Carol Smith was onto this trend more than a decade ago, when the incidence was running no higher than 1 out of every 500 children. It's now up at something more like 1 in 100 children. That's 1 percent of the population!"
Robert McClure (InvestigateWest, 02/02/2010).

Well, autism is skyrocketing all over the World where heavy manmade mutagenic pollution exists.

The diagnostic-criteria-is-expanded explanation is wrong because the diagnostic criteria have become more precise and standardized. The more-adapt-at-noticing-it explanation is also wrong because autism is an extreme disorder that cannot be overlooked, and, importantly, autism is new in the histories of families. Autistics do not have children, so here natural selection is working with maximum strength. But now we do have adult autistics too, because the autism explosion started more than 60 years ago when mutagenic pollution exploded: mutagenic car exhaust, mutagenic industrial smoke, mutagenic smoke from burning buildings (and other things), mutagenic synthetic chemicals, mutagenic manmade radiation, mutagenic medicine, mutagenic pesticides, mutagenic mother's milk (all milk contains dioxins, for example), mutagenic toys, mutagenic everything.

"Once rare, autism has reached epidemic proportions in the United States. The increase cannot be attributed to changes in diagnostic criteria, which have actually become more restrictive. Already a heavy burden on educational facilities, the increasing number of patients afflicted with this serious disability will have an enormous effect on the economy as the affected children reach adulthood. Studies of all possible causes of the epidemic are urgently needed. To date, studies of a potential relationship to childhood vaccines have been limited and flawed. [...] The important historical observation about autism is that it was unknown in ancient cultures, or even in medieval times, and that it just appeared some 60 years ago. Leo Kanner, while at Johns Hopkins, was first to describe autism in 1943. His article described 11 children who had an apparently rare syndrome of extreme autistic aloneness. Because these children's symptoms started early, Kanner's Syndrome was also known as infantile autism. In 1944, Hans Asperger also described a group of children with similar symptoms who were highly recognizable."
F. Edward Yazbak (See: http://www.jpands.org/vol8no4/yazbak.pdf)

However, note that the American Medical Association claims that "Diagnoses of autism have increased in the past few decades, but it is important to note that the increase is attributed to a broadened definition of ASDs and better recognition on the part of physicians of autism symptoms." Clearly, AMA has not yet discovered the GGC. Their claim is unfounded wishful thinking, only. (See above.)

The following quote is a copy of a message to www.aspiesforfreedom.com from Alison:

Quote:

"From New Scientist, 21 August 2010, page 4

Autism Explosion

Why have the numbers of autism diagnoses increased seven-fold in recent decades? Part of the increase could be down to a trend for later parenthood in the west, yet half of the rise remains a mystery. A series of US studies by Peter Bearman of Columbia University in New York identified changes in diagnosis as the most important factor. He estimates that a switch from diagnoses of mental retardation to autism accounts for 26 per cent of extra cases seen in California between 1993 and 2005. But a trend towards having children later in life also plays a part, since older parents are more likely to give birth to a child with autism. Together with increased awareness of the condition, this factor accounts for another 25 per cent of the increase. Environmental factors are the most likely cause of the remaining 50 per cent, although no definite candidates have been identified, says Tom Insel, director of the National Institute of Mental Health in Rockville, Maryland. "These studies give me the feeling that there must be a true increase in the number of children affected," Insel says."

Unquote.

And AMA also tells us:

"Genetics of Autism
Autism is the most common disorder in a group of neurodevelopmental disorders called the Autism Spectrum Disorders (ASD). It is characterized by impairments in social interaction, deficits in verbal and non-verbal communication, and restricted repetitive patterns of behavior and interests. ASDs also include Asperger syndrome and pervasive developmental disorder not otherwise specified (PDD-nos). Autism is estimated to affect 15-20 in 10,000 children, while all ASDs combined affect approximately 60 in 10,000 children. Autism is strongly genetically determined, as demonstrated by its increased prevalence in siblings. Monozygotic twins show 60%-90% concordance, meaning that in 60%-90% of cases in which one twin has autism, the other twin does also. Concordance in dizygotic twins and siblings is 5%-10%. Males are affected with ASDs four times as often as females." (See: www.ama-assn.org)

For a closer look at the genetics of autism, see, for example:

www.exploringautism.org/genetics/overview.htm
www.actionbioscience.org/genomic/dougherty.html

The earlier-"retarded"-is now-"autistic" explanation is wrong because autistics are not really retarded. Even Albert Einstein was, wrongly, said to be autistic. Besides, the number of retarded has increased as well, due to the GGC, of course.

The social-skills-are-more-important-today explanation is wrong because autism is much more than lack of social skills. (See below.) The social explanation is wrong because the SSSM (Standard Social Science Model) is proved to be wrong. (See, for example, *The Nurture Assumption* by Judith Rich Harris.)

"Some babies – notably those with the disorder called autism – don't do this. Autistic babies don't look their parents in the eye, don't smile at them, don't seem glad to see them. It is difficult to feel enthusiastic about a baby who isn't enthusiastic about you. It's difficult to interact with a child who won't look at you. The late Bruno Bettelheim, who for many years ran an institution for autistic children, claimed that autism was caused by the mother's coldness, her lack of feeling for the child. One of these mothers later attacked Bettelheim in print, calling him a "vile individual" who had "brought ostracism and suffering to entire families." Bettelheim was not only cruel: he was wrong. Autism is caused by a defect in the brain; autistic children are born that way. The mothers' apparent coldness was not the cause of their children's behavioral abnormalities – it was a reaction to it." (Judith Rich Harris.)

The vaccine explanation is proved to be wrong. If it was right, it would actually mean a type of genetic disease. And this disease would be just autism. You see, the autistics are gene-damaged in many ways. The American Medical Association concludes that there is no scientific evidence that a link exists between incidence of autism and vaccinations.

If the milk-protein explanation (or similar explanations) was right, it would actually mean a type of genetic disease. And these diseases would, indeed, be diseases in the autism spectrum.

Researching the families, we find that autism, in general, is earlier unknown in the history of the families. Autism is totally lacking in older history, older family histories, older fiction literature, older medicine literature, older philosophical literature, older psychological literature, and older psychiatric literature.

The autism spectrum of diseases is, I'm sorry to say, one more tragic result of the manmade global general genetic catastrophe caused by manmade local and global manmade mutagenic pollution (including radiation and dangerous noise).

11. Neglecting, boycotting, explaining away, covering up

Genetic mental disorders (genetic behavioral disorders) and genetic physical (anatomical and physiological) disorders are exploding all over the world. Just like manmade climate change was ignored (since scientists in Sweden found, in the 1800s, that the use of fossil fuel would damage Earth's climate), today manmade gene damage is, in general, ignored by today's scientists. Newspapers, magazines, TV stations, radio stations, universities, organizations, scientists, governments, politicians, and others are, in general, ignoring, boycotting, explaining away, and/or covering up the general genetic catastrophe.

I've contacted *New York Times* several times about the general genetic catastrophe. I've sent letters, e-mails, book copies -- a lot of information. No answer. No feedback. Total silence. Why? I don't know. Has it something to do with editors and journalists not willing to publish bad news about science and science-based technology?

I've contacted *Scientific American* several times about the general genetic gatastrophe. I've sent letters, e-mails, book copies -- a lot of information. No answer. No feedback. Total silence. Why? I don't know. Has it something to do with scientists, science editors, and science journalists not willing to publish bad news about science and science-based technology?

I've sent several letters (mostly as e-mails), containing a lot of information on gene damage and the general genetic catastrophe, to *Time* and *Newsweek*. Automatic reply always, saying automatically that they will come back later. But they never did. I tried to motivate them to research these problems. They received a lot of ideas. But absolute silence. What shall I say? I think *Time* and *Newsweek* are boycotting this very subject just as they probably boycotted info on manmade climatic change before the slow mainstream science finally began to talk -- perhaps too late.

Norwegian national newspapers are avoiding publishing information on genetic damage and the general genetic catastrophe. I've contacted the three largest national Norwegian newspapers: *Aftenposten*, *VG* (*Verdens Gang*), and *Dagbladet*. They do not want to publish info on this subject. The same is the case for *Klassekampen* (a national communist newspaper). Local newspapers (*Nordlys*, *Rana Blad*, etc.), exactly the same reaction. They say this is not a *local* problem!

Even Norwegian environmental organizations (such as the internationally famous *Bellona*) are boycotting info on the GGC.

And so on, and so forth.

The following list contains 20 ways to explain away the general genetic catastrophe.

1. "Business as usual. All these genetic diseases have always existed."

2. "The upward trend is not real. This is due to better diagnosis."

3. "These diseases are caused by our lifestyle."

4. "These diseases are caused by unhealthy food."

5. "These diseases are caused by viruses."

6. "These conditions are environmental."

7. "The increased lifespan disproves that we have a genetic catastrophe."

8. "The upward trend is due to increased lifespan."

9. "A genetic catastrophe is not possible because of natural selection."

10. "A genetic catastrophe is not possible because all gene-damaged fetuses are naturally aborted." [It's true that 75 percent of all fetuses are "naturally" aborted]

11. "A genetic catastrophe is not possible because the immune system kills all gene-damaged cells."

12. "A genetic catastrophe is not possible because all gene-damaged cells commit suicide (apoptosis)."

13. "A genetic catastrophe is not possible because of DNA repair."

14. "These conditions are a result of evolution."

15. "These conditions are adaptations."

16. "The explanation is that our bodies simply wear out."

17. "No, you don't have any evidence."

18. "You [Nils K. Oeijord] are simply a crackpot with an odd, paranoid 'genetic catastrophe' fixation who deserves little of our time." (Eric)

19. "The GGC is an absurd idea caused by paranoia. It is utter garbage, unsupported by evidence, unreferenced, cranky." (Alec)

20. "The scientific community publishes a very nice journal called "Human Mutation". We get it at our lab. It contains many very thorough and detailed papers regarding all aspects of human genetic diseases. It also publishes the Mutation and Polymorphism Report. Tons of good information in there as well. Our Center for Disease Control publishes a weekly Morbidity and Mortality Report. And you're trying to tell me that with the thousands of well trained and concerned people around the world reading these publications every week – not to mention the thousands of other good, relevant scientific journals – that somehow ALL of them are completely missing some absolutely catastrophic trend? (Eric)

****

Author's (Nils K. Oeijord's) anwer:

Dear Alec and Eric!

Thanks a lot, Alec and Eric! You seem to have documented above That I'm the discoverer of the General Genetic Catastrophe.

Best regards,
Nils K. Oeijord

12. Finally: Recognition

I (Nils K. Oeijord) am the discoverer of the *general genetic catastrophe due to manmade local and global mutagenic pollution*. I want me to get the recognition I deserve. So I'm happy that I have earned a place in *Who's Who in the World* (28th Edition), in *Great Minds of the 21st Century* (5th Edition), and in *2000 Outstanding Intellectuals of the 21st Century* (2011 Edition). However, I'm pessimistic that the scientists and the politicians are capable of understanding how to react to what is discovered. Remember that science and politics ignored the discovery (1859 – 1896) of global warming. And this time the whole problem is much larger and much more complex.

VII The road ahead

Technological progress is like an axe in the hands of a pathological criminal.
Albert Einstein

"However beautiful the strategy, you should occasionally look at the results."
Winston Churchill

"Some drugs can cause permanent changes to the DNA of germ cells - egg cells and sperm cells - leading to mutations which are inherited by a patient's children. One example is nitrogen mustard, a compound similar to mustard gas, which is used to destroy cancer cells in chemotherapy. The drug destroys cancer cells by linking their DNA strands together with a nitrogen atom, but can also link the DNA of germ cells, leading to deletions of DNA bases, thereby causing mutations. For this reason, known mutagens are usually only used deliberately to kill cancer cells."

BBC Homepage

The indirect effects of the GGC are terrible: war, uprising, criminality, violence, murder, and hunger. The potential for human chaos is unlimited. It's easy to see that the GGC interacts with the CC (Climatic Catastrophe). The CC is much more damaging than mainstream science believes today. The methane molecule (CH_4) has 23 times larger greenhouse effect than the CO_2 molecule. From the continental shelves and the gulfs of the world's oceans methane is increasingly leaking out because of warmer oceans and warmer seafloors, and because of cracks due to oil and gas production. The permafrost soil contains twice as much carbon as the total carbon in today's atmosphere. Now the permafrost is starting to melt and is increasingly releasing CO2 and methane into the atmosphere. Additionally, catastrophic amounts of methane are released from the world's enormous areas of rice fields, from some 1 billion cows, and from enormous areas of humanmade termite ecosystems. The oceans have absorbed ca. 50 percent of humanmade CO2 pollution, but now the warmer seawater has decreased CO2 absorbing capacity. On the contrary, the oceans may soon start to release CO2 back into the atmosphere. And, as we already know, catastrophic lower pH in the oceans caused by human CO2 pollution has already started the killing of ocean organisms. What happens if the oceans die? What happens if the oxygen-producing algae in the oceans are killed? How much deadly H2S is then produced? Science is silent.

High CO2 concentration will destroy the balance between the green vegetation and plant eating insects, for example. Trees in the rainforests, for example, can be deadly attacked, creating dry organic materials causing the rainforests to burn down. The world's rainforests contain about five times the amount of CO2 that is already in the atmosphere.

I cannot see how the untalented and uninterested scientists and politicians of the world are able to avoid all these scenarios.

Therefore, I believe that the humanmade contribution to the *increase* of the average temperature of the surface-near atmosphere by 2100 will be 10 – 15 degrees Celsius. I think many researchers agree with me, but they dare not tell us what they think. And they dare not honestly research this problem, either.

It certainly is obvious why it is important to protect our DNA. We must have a speech to the Congress about the GGC as soon as possible. We must have lectures on the GGC at the colleges and the universities. However, again, I'm pessimistic that the scientists and the politicians are capable of understanding how to react to what is discovered. James Hansen's speech to Congress on June 23, 1988, was a seminal moment in the global warming debate. But since then not much has happened. On the contrary: pollution by climate gases has increased enormously since 1988.

I've an idea. Let's try *competition*: Competition between individuals, competition between teams, competition between regions, competition between countries. Competition not measured in units of money, but units of pollution. Thanks to competition we landed on the Moon.

VIII Readings/Study
(with a list of web search words)

"The Theory of Evolution is being invoked frequently in medicine to account for counter-intuitive findings such as programmed cell death in congestive heart failure and the fact that fever can prove both blessing or curse, depending on the circumstances. These and other examples of what might be called maladaptive adaptations are discussed, and it is suggested that human development may have reached a stage where the roles of mutation and selection are drastically changed. By creating an environment both mutagenic and protective, we have altered the balance between the two great driving forces of evolution, increasing the frequency of mutations and reducing the need for adaptation. As a result, new diseases have arisen and the whole evolutionary process seems to have lost some of its benevolence, no longer insuring the survival of the fittest. We must entertain the possibility that Darwin's theory cannot explain the last few millennia of human evolution and is now useful chiefly as an approximation applicable to very long periods of time."

J. Herman (1996)

"Mental health problems do not affect three or four out of every five persons but one out of one."

William Menninger

"I have suffered from depression for most of my life. It is an illness."

Ada Ant

"All" of us know the DSM-IV-TR (Diagnostic and Statistical Manual of Mental Disorders.) But what is DSM-IV-TR, basically? The index does not contain words like "gene damage", "genetic damage", "genetic disease", or "genetic disorder". I do not think the authors of this book do really fully understand what they are writing about. The fact is that

DSM-IV-TR is a catalogue of genetic damage to the human instincts (behavior modules). Undoubtedly, basically the human mental diseases are gene-damaged human instincts. (A long list of genetic damage in humans is found in my book *Derailed Evolution*.)

Dogs and humans get many of the same genetic diseases. Researchers are beginning to use dogs as models of genetic psychiatric diseases and genetic behavioral diseases, such as, for example, obsessive-compulsive disorder (OCD) and autism. (A long list of genetic damage in dogs is found in my book *Derailed Evolution*.) By the way: some 50 percent of all dogs in the USA die of cancer.

"Open any catalogue of the human genome and you will be confronted not with a list of human potentialities, but with a list of diseases, mostly ones named after pairs of obscure central-European doctors. This gene causes Niemann-Pick disease; that one causes Wolf-Hirschhorn syndrome. The impression given is that genes are there to cause diseases. 'New gene for mental illness,' announces a website on genes that reports the latest news from the front, 'The gene for early-onset dystonia. Gene for kidney cancer. Gene for kidney cancer isolated. Autism linked to serotonin transporter gene. A new Alzheimer's gene. The genetics of obsessive behavior.'" (Matt Ridley: *Genome*.) Well, it is true that the only thing we know about some genes is that their malfunction causes a special disease.

Gene mapping is the process of establishing the locations of genes on the chromosomes.
Gene mapping, see, for example: www.genome.gov/10000715

A gene atlas identifies what genes are being expressed in different tissues.

Some gene (expression) atlas:

www.geneatlas.org/gene/;jsessionid=1676861294511423629

http://genatlas.medecine.univ-paris5.fr/

www.fimm.fi/en/scientific_highlights/the_new_finnish
_gene_atlas_places_finns_on_the_world-s_genetic_map/

www.ebi.ac.uk/microarray/doc/atlas/index.html

Human protein atlas: www.proteinatlas.org/

Cancer genome atlas: http://cancergenome.nih.gov/

We must understand the past on its own terms without any reference to "what happened next." A new generation of leading historians of science has shown that seemingly crucial scientific experiments were often fatally flawed and that results were often modified to suit the case being argued. Evidence is now available to show that many leading scientists used political influence to advance their cause.

"The great biologist Louis Pasteur suppressed data that didn't support the case he was making. Albert Einstein's theory of general relativity was only 'confirmed' in 1919 because an eminent British scientist messaged his figures. Joseph Lister's famously spotless hospital wards were actually notoriously dirty. Gregor Mendel, supposed father of the science of heredity, never grasped the fundamental principles of 'Mendelian' genetics." (John Waller: *Fabulous science*.)

Science is a field filled with propaganda and fundamentalism, that is: *Science represents progress and Science represents the only truth*. And Science is a God. Even John Waller writes about "serious sins against science." Dawkins' equation: Evolution (or science) exists = God does not exist.

Gregor Mendel's epoch-making paper was first presented in 1865, but was discovered and understood only in the early 1900s. Mendel never grasped the basic tenets of Mendelian genetics. And Darwin ignored Francis Galton's ingenious particulate theory of heredity (available in the 1860s), but "his" theory of blending inheritance was to cause him immense discomfort. The word "gene" was not coined until 1903. Chance mutations were first studied in the 1910s. In the 1930s it was fully realized that Darwinism and Mendelism made a fine matching pair. But even today - 75 years later – science does not understand that evolutionary theories are crucial to the future of the human populations. This sad situation partly explains why science still does not understand that we have a general genetic catastrophe. Of course, scientific products, by-products, activities, and results damaged our genes even before 1903, but before 1903 science did not know the words "gene damage." This fact does illustrate how extremely dangerous the totality of science is. It does show that science is blind when it comes to the real big things. Karl Popper made a useful distinction between *discovery* and *verification* in the development of knowledge. Science only becomes reliable knowledge, Popper argued in 1959, after its validity has been tested over the course of many years. Popper was "inclined to think that scientific discovery is impossible without faith in ideas which are of a purely speculative kind, and sometimes even quite hazy." Certainly, polluting the whole world (including our bodies) with mutagenic and carcinogenic chemicals and radiation is quite hazy.

The *Bibliography* (*Books*) of this book contains a large number of important books. Sadly, these kinds of books have very few buyers. People (including politicians and scientists) do not want to learn about these things. Therefore, the terrible facts described, explained, and published in these books are known to few.

Likewise, the websites in *A list of websites*, in this book, contain dramatic information known to few. But hundreds of other websites have published equally important information.

The author's *Google groups* and *Yahoo! groups* (see section IV) are perhaps useful places to start the study of the general genetic catastrophe.

Here follows a list of *web search words* useful for our purpose:

A

abnormal
Abnormal allele
Abnormal alleles
Abnormal chromosome
Abnormal chromosomes
Abnormal gene
Abnormal genes
Abnormal genome
Abnormal genomes
Abnormality
Acquired genetic disease
Acquired genetic diseases
Act abnormally
Acts abnormally
Adaptive/ inducible repair
Add genetic material
Addition
Addition mutation
Addition mutations
Alkylate DNA
Alkylating agent
Alkylating agents
Allele
Alleles
Alteration of a DNA sequence
Altered chromosome
Altered chromosomes
Altered DNA
Altered DNA sequences
Altered gene
Altered genes
Altered protein
Altered proteins
Ames test
Apoptosis
A-specific mutagen
A-specific mutagens

B

Base analog
Base analogs
Base excision repair
Base modification
Base substitution mutagen
Base substitution mutagens
Base substitution mutation
Base substitution mutations
Because of damaged genes
Bind to DNA
Biological effect
Biological effects
Birth defect
Birth defects
Birth rate
Birth rates

C

Cancer
Cancer fumes
Carcinogen
Carcinogenesis
Carcinogenity
Carcinogens
Cause genetic catastrophe
Causes genetic catastrophe
Cause genetic disaster
Causes genetic disaster
Cell phone radiation
Cell phone radiation cancer
Cells act abnormally
Cell-suicide mechanism
Cellular phone radiation

Changed chromosome
Changed chromosomes
Changed DNA
Changed gene
Changed genes
Changed genome
Changed genomes
Changes in chromosome structure
Changes in mRNA structure
Changes in single DNA nucleotides
Checkpoint against gene damage
Checkpoint pathway
Chemical damage to genes
Chromosomal damage
Chromosomal mutation
Chromosomal mutations
Chromosome aberration
Chromosome aberrations
Chromosome abnormalities
Chromosome abnormality
Chromosome break
Chromosome breaks
Chromosome damage
Chromosome deletion
Chromosome deletions
Chromosome disease
Chromosome diseases
Chromosome disorder
Chromosome disorders
Chromosome error
Chromosome error statistics
Chromosome errors
Chromosome illness
Chromosome illnesses
Chromosome mutation
Chromosome mutations
Chromosome protecting function
Chromosome repair
Cleanup operation
Cleanup operations
Common genetic disease
Common genetic diseases

Condition
Conditions
Covalent lesion
Covalent lesions
Critical genes
C-specific mutagen
C-specific mutagens

D

Damage gene
Damaged gene
Damaged genes
Damage to DNA
Damage to chromosomes
Damage to genes
Damage to the genome
Damaged allele
Damaged alleles
Damaged chromosome
Damaged chromosomes
Damaged gene
Damaged genes
Damaged genome
Damaged genomes
Damaged instinct
Damaged instincts
Dangers of science
Dangers of technology
Defect
Defective cell
Deficiency
Delete genetic material
Deletion
Deletion mutation
Deletion mutations
Deletion of a gene
Deletion of genes
Deletions
Derailed evolution

Frameshift mutation
Frameshift mutations
Frequencies of mutations
Frequency of inherited disorders
Fungicides cause gene damage
Fungicides cause genetic damage

G

Gene
Gene alteration
Gene damage
Gene damage prevention
Gene-damaged
Gene deletion
Gene deletions
Gene disorder
Gene duplication
Gene error
Gene error statistics
Gene errors
Gene expression
Gene expressions
Gene function
Gene instability
Gene modification
Gene stability
Generate mutation
Generate mutations
Genes
Genes and cancer
Genes predisposing
Genes that repair damaged genes
Genetic alteration
Genetic alterations
Genetic birth defect
Genetic birth defects
Genetic birth malformation
Genetic birth malformations
Genetic catastrophe

Genetic cause
Genetic change
Genetic changes
Genetic chaos
Genetic condition
Genetic conditions
Genetic damage
Genetic damage prevention
Genetic determination
Genetic disabilities
Genetic disability
Genetic disaster
Genetic disease
Genetic diseases
Genetic disorder
Genetic disorders
Genetic effect
Genetic effects
Genetic epidemic
Genetic epidemics
Genetic error
Genetic error statistics
Genetic errors
Genetic failure
Genetic future
Genetic handicap
Genetic heart disease
Genetic illness
Genetic instability
Genetic malformations
Genetic manipulation
Genetic manipulations
Genetic modification
Genetic modifications
Genetic mutation
Genetic mutations
Genetic origins of disease
Genetic origins of diseases
Genetic risk
Genetic syndrome
Genetic syndromes
Genetic threat

Genetic toxicology
Genetic tragedy
Genetically caused
Genetically determined
Genome
Genome alteration
Genome damage
Genome error
Genome errors
Genomes
Genome damage
Genome duplication
Genome instability
Genome stability
Genomic
Genomic instability
Genotoxic
Genotoxic agent
Genotoxic agents
Genotoxic carcinogen
Genotoxic carcinogens
Genotoxic chemical
Genotoxic chemicals
Genotoxic compound
Genotoxic compounds
Genotoxic damage
Genotoxic effect
Genotoxic effects
Genotoxic impurities
Genotoxic impurity
Genotoxic pollutant
Genotoxic pollutants
Genotoxic pollution
Genotoxic pollutions
Genotoxicity
Germinal mutation
Germinal mutations
Germline mutation
Germline mutations
G-specific mutagen
G-specific mutagens

H

Handicap
Hayflick Limit
Hazards of electromagnetic radiation
Hazards of radiation
Hazards of radioactivity
Health hazard
Health hazards
Herbicides cause gene damage
Herbicides cause genetic damage
Hereditary
Hereditary disease
Hereditary diseases
Hereditary disorder
Hereditary disorders
Hereditary illness
Hereditary illnesses
Hereditary mutation
Hereditary mutations
Heritable chromosome mutation
Heritable chromosome mutations
Heritable disease
Heritable diseases
Heritable disorder
Heritable disorders
Heritable DNA mutation
Heritable DNA mutations
Heritable illness
Heritable illnesses
Heritable mutation
Heritable mutations
Heritable gene damage
How do mutations occur
How genes get damaged
How to avoid gene damage
How to avoid genetic damage
Human gene
Human genes
Human genome

I

Illness
Illnesses
Implication of gene damage
Implication of genetic damage
Implications of gene damage
Implications of genetic damage
Inborn birth malformation
Inborn birth malformations
Inborn disease
Inborn diseases
Inborn disorder
Inborn disorders
Inborn errors (of ...)
Induce mutation
Induce mutations
Induced mutation
Induced mutations
Inherited birth malformation
Inherited birth malformations
Inherited disease
Inherited diseases
Inherited disorder
Inherited disorders
Inherited genetic disease
Inherited genetic diseases
Inherited rare disease
Inherited rare diseases
Insertion
Insertion mutation
Insertion mutations
Instanility
Instability of a duplication
Instability of a tandem duplication
Instability of chromosome
Instability of chromosomes
Instability of DNA
Instability of gene
Instability of genes
Integrity of the genome
Interact with DNA

Inversion
Inversion of genes
Ionizing radiation
Irradiation

J

Junk DNA
Junk fragments
Junk fragments of damaged DNA

L

Lesion
Lesion-binding protein
Lesion-binding proteins
Lesion persistence
Lesion recognition
Lesions
Level of ionizing radiation
Level of radiation
Levels of ionizing radiation
Levels of radiation
Ligase
Ligases
List of gene damage
List of genetic conditions
List of genetic damage
List of genetic diseases
List of genetic disorders
List of genetic illnesses
List of mutations
Long term exposure
Loss of tandem duplication
Loss of various plasmid types
Low dose

M

Mammograms damage genes
Medical genetics
Mental disease
Mental diseases
Mental disorder
Mental disorders
Mental illness
Mental illnesses
Microwaves
Mismatch
Mismatch repair
Mismatched bases
Missense mutation
Missense mutations
Mitochondrial DNA
Mitochondrial DNA mutation
Mitochondrial DNA mutation rate
Mitochondrial DNA mutation rates
Medicines cause gene damage
Medicines cause genetic damage
Mobile phone radiation
Modified gene
Modified genes
Modifying genes
Mortality rate
Mortality rates
Mutagen
Mutagenesis
Mutagen's target specificity
Mutagenic activity
Mutagenic additive
Mutagenic additives
Mutagenic agent
Mutagenic agents
Mutagenic air pollution
Mutagenic asbestos
Mutagenic chemicals
Mutagenic drug
Mutagenic drugs
Mutagenic dye

Mutagenic dyes
Mutagenic gasoline vapor
Mutagenic heavy metals
Mutagenic pesticide
Mutagenic pesticides
Mutagenic radiation
Mutagenic solvent
Mutagenic solvents
Mutagenity
Mutagens
Mutant
Mutants
Mutated allele
Mutated alleles
Mutated chromosome
Mutated chromosomes
Mutated gene
Mutated genes
Mutation
Mutation carrier
Mutation carriers
Mutation in a translated region
Mutation in an untranslated region
Mutation rate
Mutation rates
Mutational
Mutational mechanism
Mutations
Mutations in the DNA
Mutations in the human genome
Mutations in translated regions
Mutations in untranslated regions
Mutator gene
Mutator genes

N

Natural mutation frequency
Natural mutations
Non-disjunction

Non-disjunction mutation
Non-disjunction mutations
Non-duplicated
Non-duplicated chromosomes
Non-duplicated genes
Nonsense mutation
Nonsense mutations
Non-wild-type mutation
Non-wild-type mutations
Normal chromosome
Normal chromosomes
Normal DNA
Normal gene
Normal genes
Normal repair process
Normal repair processes
Niclear weapons tests
Nucleic acid
Nucleotide excision repair

O

Oncogen
Oncogene
Oncogenes
Orphan disease
Orphan diseases
Our future
Our stolen future
Over generations
Oncogens

P

Persistent genomic instability
Pesticides cause gene damage
Pesticides cause genetic damage

Phone mast radiation
Photoreactivation
Physical change in a chromosome
Physical change in a gene
Physical disease
Physical diseases
Physical disorder
Physical disorders
Physical illness
Physical illnesses
Point mutation
Point mutations
Pollute
Polluted
Possible carcinogen
Possible carcinogens
Possible mutagen
Possible mutagens
Post-replication repair
Precautionary principle
Predisposing
Prevention of gene damage
Prevention of genetic damage
Protein-DNA lesion-recognition
Protooncogen
Protooncogens
p53-gene

R

Radar
Radar radiation
Radiation limits
Radiation medicine
Radioactive contamination
Radioactive contaminations
Radiological
Radiological protection
Radiation biology
Random chromosome damage

T-specific mutagen
T-specific mutagens
Tumor suppressor gene
Tumor suppressor genes
Types of gene damage
Types of genetic damage
Types of mutations

U

Unnatural selection
Unstable chromosomes
Unstable DNA
Unstable genes
U-specific mutagen
U-specific mutagens
UV damaged gene
UV damaged genes
Unstable gene

W

Whole genome duplication
Wild-type mutation
Wild-type mutations

X

X-rays

Y

Y chromosome
Y chromosome can repair its own genes

Appendix A:
Some basic genetics

"Natural selection can do nothing until favorable individual differences or variations occur."
C. Darwin

"The faithful duplication and repair exhibited by the double-stranded DNA structure would seem to be incompatible with the process of evolution. Therefore, evolution has been explained by the occurrence of errors during DNA replication and repair."
Mainstream science

"It is a considerable strain on one's credulity to assume that finely balanced systems such as certain sense organs (the eye of vertebrates, or the bird's feather) could be improved by random mutations. This is even more true of some ecological chain relationships. However, objectors to random mutations have so far been unable to advance any alternative explanation that was supported by substantial evidence."
Harvard biologist Ernst Mayr

"The real difficulty of Darwinism is the well-known problem of explaining an evolution which prima facie may look goal-directed, such as that of our eyes, by an incredibly large number of very small steps; for according to Darwinism, each of these steps is the result of a purely accidental mutation. That all these independent accidental mutations should have had survival value is difficult to explain."
Sir Karl Popper, widely regarded as the foremost philosopher of science

"To postulate that the development and survival of the fittest is entirely a consequence of chance mutations seems to me a hypothesis based on no evidence and irreconcilable with the facts. These classical evolutionary theories are a gross oversimplification of an immensely complex and intricate mass of facts, and it amazes me that they are swallowed so uncritically and readily, and for such a long time, by so many scientists without a murmur of protest."
Ernst Chain, biologist who won a Nobel Prize for penicillin research.

The human **genome** is the entirety of a human's hereditary material, and consists of twenty-three pairs of **chromosmes** (see below). One pair is the sex chromosomes. Women have two large X chromosomes. Men have one (large) X chromosome and one small Y chromosome.

DNA is the hereditary material in living organisms. A **gene** is a unit of heredity in a living organism, and is a stretch of a DNA molecule that codes for a type of protein. One gene codes for the production of one protein. In each cell, the DNA molecule is packaged into thread-like structures called **chromosomes**. A **mutation** is a permanent change in the hereditary material. Mutation in a gene can alter the protein encoded by the gene. Chromosome mutations are much more dramatic than gene mutations.

The human genome contains approximately three billion **base pairs**. Base pairs are pairs of **nucleotides** (in the DNA molecule) joined with a hydrogen bond. The nucleotides are adenine (A), thymine (T), cytosine (C), and guanine (G). (In RNA thymine is replaced with uracil (U).) The only possible base pairs in DNA are AT, TA, CG, and GC. A **point mutation** is when a single base pair is altered.

Some characteristics come from a single gene, whereas others come from gene combinations. Humans have some 30,000 coding genes. (See below.)

"In biology, a **mutagen** (Latin, literally *origin of change*) is a physical or chemical agent that changes the genetic material, usually DNA, of an organism and thus increases the frequency of mutations above the natural background level. As many mutations cause cancer, mutagens are typically also carcinogens. Not all mutations are caused by mutagens: so-called "spontaneous mutations" occur due to errors in DNA replication, repair and recombination." (Wikipedia.)

"Mutations that are caused by agents that damage the DNA are known as **induced mutations**. Agents that mutate DNA are called **mutagens** and are of three main types: mutagenic chemicals, radiation and heat. Even if there are no dangerous chemicals or radiation around, mutations still occur, though less frequently. These are **spontaneous mutations**. Some of these are due to errors in DNA replication. The enzymes of DNA replication are not perfect and sometimes make mistakes. In addition, DNA undergoes certain spontaneous chemical reactions (alterations) at a low but detectable rate and this rate goes up with increasing temperature." (David P. Clark.)

"Recombination, repair, and replication (the 3-R's) are intimately associated with many types of mutations and chromosome rearrangements. CSG investigations of the 3-R's in yeast have led to the development of several new approaches and concepts including direct RNA repair of DNA double-strand breaks (DSBs), ability to measure resection at random DSBs, mechanisms and consequences of altered lagging strand replication, as well as hypermutability by environmental agents at DSB resected ends."(The Chromosome Stability Group (CSG).)

"Mutagenic pollution of the natural environment is undoubtedly a serious and general problem. Reports of various agencies indicate that the presence of mutagenic compounds in different habitats is a common phenomenon rather than an exception (see, for e.g. Davey 1999). The list of known mutagenic chemicals is very long. The Environmental Mutagen Information Center database (http://www.nlm.nih.gov/pubs/factsheets/emicfs.html) contains over 20 000 citations to literature on agents that have been tested for mutagenic activity. The problem of the presence of mutagenic chemicals in natural habitats is very important because such compounds are capable of inducing serious diseases, including cancer. Moreover, they can potentially damage the germ line of higher organisms, which may lead to fertility problems and to negative genetic changes in future generations (reviewed by Mortelmans and Zeiger 2000). Therefore, detection of mutagens in the natural environment is very important. As chemical mutagens elicit deleterious effects on living organisms at extremely low concentrations, their detection in natural habitats, at levels that can be dangerous for animals or humans, may be difficult and complicated." (G. Wegrzyn and A. Czyz)

"**Abstract**. Wild rodents *(Mus domesticus)* were collected in three areas in Rome exposed to different traffic flows to ascertain a possible correlation between genetic damage and heavy metal concentration. The concentration of lead, cadmium and zinc were determined in liver, kidney and bones and two mutagenicity tests (micronucleus test and sperm abnormality assay) were employed. The results obtained showed that the contents of lead and cadmium were higher in animals collected in areas with high traffic flows than in those from control areas. A statistically significant increase of the frequency of micronucleated erythrocytes and of abnormal sperm cells was also obtained in animals collected in sites with high traffic flows. The investigation confirmed the suitability of using wild rodents as bioindicators of environmental pollution and as key-organisms in programs of pollution monitoring and environmental conservation." (Environmental Pollution. Volume 92, Issue 3, 1996, Pages 323 -328)

"Titanium dioxide (TiO_2) nanoparticles - ubiquitous in common household products such as cosmetics, sunscreen and vitamins - cause systemic genetic damage in mice, according to a new study by researchers at UCLA's Jonsson Comprehensive Cancer Center. Senior author of the study, Robert Schiestl, said the nanoparticles induced single- and double-strand DNA breaks and also caused chromosomal damage as well as inflammation, all of which increase the risk for cancer. The UCLA study is the first to show that the nanoparticles had such an effect. [...] The manufacture of TiO_2 nanoparticles is a huge industry, with production at about two million tons per year. In addition to paint, cosmetics, sunscreen and vitamins, the nanoparticles can be found in toothpaste, food colorants, nutritional supplements and hundreds of other personal care products." (Kate Melville.)

According to a recent paper in *Science* a new human individual is born with some 70 new mutations (the same mutations in all cells), compared to the DNA of its parents. The fraction of the DNA that actually is genes in the traditional sense of making proteins is as low as ca. 1.2 percent. So ca. 98.8 percent of human DNA is so called *junk DNA*. (But not all our junk DNA is junk. It turns out that junk DNA contains stretches of DNA that have been remarkably conserved during evolution. Some of these regions have accumulated fewer mutations than encoding genes have. This suggests that these sequences are important to the organism, *but why is as yet unknown.*) 70 x 0.012 = 0.84, so: *Each new human being represents ca. 1 ordinary mutation. We know that almost all mutations observed have been harmful to the organism.* What is the success rate of mutations? We don't know. But we know that a pathetically low fraction of mutations is successful. Googling "success rate of mutations" gives 8 hits, googling "rate of good mutations" gives 18 hits, googling "rate of adaptive mutations per genome per generation" gives 5 hits. *But these 31 hits are not about the real thing. Science avoids this question.* Also: *Science is, in general, neglecting the general genetic catastrophe due to global manmade mutagenic pollution.*

Each unsuccessful mutation leads ultimately to one "genetic death", since each mutation can be eliminated only by death or failure to reproduce. (But, of course, several mutations may be picked off in the same victim.) Example (manmade selection): A screening policy (prenatal screening and abortion) intended to reduce the incidence of thalassemia exists in both jurisdictions on the island of Cyprus. Since the program's implementation in the 1970s, it has reduced the ratio of children born with thalassemia from 1 out of every 158 births to almost zero. Tests for the thalassemia gene are compulsory for both partners, prior to wedding.

The theory of random mutations and more or less random natural selection cannot explain how unsuccessful mutations are picked off. But there are a lot of bad speculation and bad mathematics found in research papers today. The most common speculative "explanations" are called *soft selection, selection early in development, and quasi-truncation selection.*

"The Human Genome Project has identified 30,000 genes and their sequence variants across different individuals. However, it leaves completely unanswered how these different genes interact to form the molecular machines that run the cell and govern its various responses. We now have the parts list, but we also need to understand the network connecting all of these parts, and how to fix it during disease." (Trey Ideker.)

The mitochondria are constituents of the cell that have their own piece of DNA. Both sons and daughters inherit their mitochondria and mitochondrial DNA *only* from their mothers. Several diseases arise from gene damage to mitochondrial DNA.

Damage to DNA-repair genes is, of course, catastrophic. Here follows three examples:

Ataxia telangiectasia is a genetic disease due to damage to a gene that makes a protein that recognizes DNA damage and blocks cell division until the damage is repaired.

Xeroderma pigmentosum is a genetic disease due to damage to a gene that makes an enzyme for cutting out damaged DNA.

The cause for hereditary nonpolyposis colon cancer (HNPCC) is due to damage to a gene that repairs DNA. At least 5 such damaged genes are found.

Appendix B:
Behavior versus action(s)
The Bronston heritability coefficient

"If liberty means anything at all, it means the right to tell people what they do not want to hear."
George Orwell

Hume's fork: Either our actions are determined, in which case we are not responsible for them, or they are the result of random events, in which case we are not responsible for them.
Oxford Dictionary of Philosophy.

Do not confuse *behavior* and *action(s)*. Yes, I know that psychology textbooks are using these words interchangingly, but they are mistaken. You know, language is genetic, but Norwegian is not genetic. Religion is genetic, but Catholicism is not genetic. Language and religion are genetic (species specific behavior), but Norwegian and Catholicism are actions, and thereby not genetic. A (genetic) alcoholic does not necessarily drink alcohol, while a (genetic) non-alcoholic may drink alcohol. And a (genetic) homosexual man may have sex with a woman. So *behavior* and *action* are totally different biological categories.

An old traditional question is: What is more important in determining an individual's *actions*, his/her genes or his/her environment? However, they who asked this question did systematically, and always, confuse *behavior* and *action*, thereby rendering the question pretty meaningless or extremely ill defined.

Traditional behavioral genetics answers the above question by computing the broad heritability coefficient. This coefficient is a statistical measure of the genetic "contribution" to differences among individuals. It tells us "what proportion of the individual differences in a population can be ascribed to genes." For example, if we say that a specific "behavior" (the geneticists confuse behavior and action) is 50 percent heritable (the coefficient = 0.50), we are saying that 50 percent of the variance in that "behavior" (see above) is "linked to heredity." The heritability coefficient of a "behavioral" trait can vary from one environment to the next because instincts are genetically situational. Note that the heritability coefficient tells you nothing about what causes the specific behavior of a specific individual. It's important to keep this in mind. And note that the heritability coefficient of complex adaptations is usually low, not high, because their genetic basis is universal and species-typical (something all of us have). This result is, of course, flawed. As I said above, the heritability coefficient does not explain what constitutes a given individual's behavioral trait. Even if the heritability (coefficient) is 50 percent (= 0.50), genes can constitute 99 percent of the trait. Moreover, the other 50 percent of the variance need not be caused by the environment, but by the measurements (for example: the researchers measure on form/off form instead of genetic differences). This is obviously the case when some researchers say that intelligence is only 60 percent heritable. Intelligence is a complex adaptation, so these results are doubly flawed (see above). Moreover, intelligence is a group of learning instincts (all instincts are learning machines; only instincts can learn; even reflexes ("little instincts") can learn), and intelligent people tend to seek out an "intelligent environment", and vice versa. More flawed! In a sense, environments are, in general, multi-genetic! So heritability coefficients are, in general, multi-flawed.

Fortunately, the problems above were solved by Mitch C. Bronston in May 2001 by asking the question "Can we quantify how well instincts learn?" His solution is to use the (broad) heritability coefficient , but let the "environmental" part of the variance quantify how well an instinct can learn. Obviously, an instinct has two innate "parts" (the "basic" part, and the learning part) and, naturally, the absolute and relative "size" of these parts are innate and vary from individual to individual, and from instinct to instinct. Also, of course, the speed of the biological development and maturation of these parts are innate, and vary from individual to individual, and from instinct to instinct. Moreover, instincts are situational, and the situationality itself is genetic! Even though we clock up more unique experiences as we age, evidence amassed over the past eighty years suggests that the "genetic contribution" to mental achievement and emotional characteristics increase with age. Example: The broad heritability coefficient of IQ is about 0.4 when measured in children, about 0.6 in adolescents, and about 0.8 in later maturity, and about 1.0 in old age. More learning causes more genetic determinism! The environmentalists are staring in incomprehension at these figures, but there is nothing inconceivable about them. The Bronston heritability coefficient tells us that the intelligence instincts' genetically determined specific learning capacities are decreasing with age. But, fortunately, this coefficient also tells us that the other part of the *total* intelligence is increasing, so that the total intelligence (is genetically determined!) is pretty constant (" = 1.0 = 100%") with age. (Individual IQ levels tend to remain unchanged from adolescence onward.) Of course, attempts to raise IQ (permanently) have failed. Both in the US and the USSR Head Start programmes that aimed to raise intelligence itself permanently totally failed. (Note: IQ tests measure action(s) and behavior(s) simultaneously.) One of the most important findings of behavior genetics has been that, statistically speaking, family environment plays no consistent role in determining personality and intelligence.

Even if environmentalism was only partly correct, the Bronston heritability coefficient had to *decrease* with age. Clearly, the *increasing* Bronston heritability coefficient for human behavioral traits proves that human behavior is instinctive, and is not created by the chaotic and poverty-stricken environmental factors. But remember: 1) all instincts are learning instincts, and 2) all instincts are situational. Obviously, human intelligence(s) and intellect(s) depend on our having more instincts, not fewer, than similar species.

The old broad heritability coefficient confuses *actions* with *behavior*. But the Bronston heritability coefficient (BHC) does not confuse actions with behavior, because, by definition, *the BHC measures how well a behavioral module (instinct) can learn.* Note that when we use the BHC, the practical measurements must be carried out in a way that does not confuse behavior and actions. The old broad heritability coefficient does not "understand" age, development, maturation, activity, and situation. But the BHC measures how age, development, maturation, activity, and situation change how well a behavioral module can learn. *The BHC understands that how well a behavioral module can learn is genetic.* The BHC says that learning capacity itself is innate, directly or indirectly, while the old broad heritability coefficient says that learning itself is mysteriously and purely environmental. Example: I (NKO) am really totally unmusical, but psychologists and educators always say to me: "No, you can learn to be musical." But I know better. The point is: Learning (as behavior) is genetic. But, of course, learning is situationally triggered. However, situationality and triggering need specific mechanisms in order to exist at all. I do not have the genetically determined musicality mechanism.

Obviously, it is meaningless to put percentages on the "contributions made by genes (or instincts) and the environment", as the old broad heritability coefficient does. Take an example: A person born with "music genes" will seek out a musical environment. Practicing in that environment (a group of musical people) will switch on certain genes to create links between certain neuropathways. These links will make the person even more musical. Altered genetic activity because of environmental factors is therefore, in a real sense, an indirect effect of genes. Human genes, human behavior, and human actions can only be understood in the context of the surroundings. Nothing in behavioral science makes sense except in the light of evolution. The smile itself is an instinct. The *action* of a smile is an *event* triggered by the environment (including our brain and our body), but the smile as *human* behavior (the smile instinct) is determined by the genes. Whether we do smile or not, *and* exactly how we smile or exactly how we do not smile, in a certain situation, are determined by how the genetically determined smile instinct, in the specific person, does learn, and *that* is basically a genetic thing. Remember: All human instincts are, more or less, capable of (instinctual) learning. We all know that from everyday experiences. We all do understand that genes do not determine *actions* directly. But genes determine learning instincts. And instincts determine (potential) behavior. Finally, (potential) behavior and environmental triggering determine actions. Again, don't confuse behavior and action as traditional psychology and traditional behavioral genetics do. The choice of X may be culturally triggered, but the deeper reasons for choosing X do reflect an instinctual process. "Good" genes (or instincts) exploit a "good" environment much better than "bad" genes (or instincts). The difference between good (high-yielding) and bad (low-yielding) plant varieties is much bigger on good soil than on bad soil. A good environment causes more genetic inequality. As a general rule, as environments become more uniform the broad heritability coefficient rises (= the "genetic" part rises). How is it possible? Because the "environmental" part is *not* environmental! It's genetic! More environmental equality causes more genetic inequality.

Appendix C:
"All diseases are genetic"

"The Human Genome Project that deciphered the human genetic code, uncovered thousands of genes that, if mutated, are involved in human genetic diseases. The genomes of many other organisms were deciphered in parallel. This now allows the evolution of these disease associated genes to be systematically studied."
Science Daily, Oct. 19, 2008.
[Author's comment: *All* coding genes are disease associated genes because *all* genes can be damaged by mutagens, high temperature, etc.]

Nobel-laureate and co-inventor of genetic engineering, Paul Berg, said that all disease is genetic even when it is something else. Here are some dramatic examples with explanations:

"Infectious diseases introduced with Europeans, like smallpox and measles, spread from one Indian tribe to another, far in advance of Europeans themselves, and killed an estimated 95 % of the New World's population."
Jared Diamond.

"It's striking that Native Americans evolved no devasting epidemic diseases to give to Europeans, in return for the many devasting epidemic diseases that Indians received from the Old World."
Jared Diamond.

"Measles and TB evolved from diseases of our cattle, influenza from a disease of pigs, and smallpox possibly from a disease of camels. The Americas had very few native domesticated animal species from which humans could acquire such diseases."
Jared Diamond.

"Genetic resistance to AIDS works in different ways and appears in different ethnic groups. The most powerful form of resistance, caused by a genetic defect, is limited to people with European or Central Asian heritage. An estimated 1 percent of people descended from Northern Europeans are virtually immune to AIDS infection, with Swedes the most likely to be protected. One theory suggests that the mutation developed in Scandinavia and moved southward with Vikings raiders."
Randy Dotinga, Med-Tech Health.

"Two hypotheses have been suggested to account for increased immunity in progeny of survivors of a disease. One hypothesis is that the progeny inherit genetic factors which make their parents resistant. The other hypothesis is that the progeny acquire an immunity either active or passive. […] The continued decline in death rates in later generations favors the genetic hypothesis. A steady increase in survival value would be expected from continued selection of genetic factors for resistance, but it is not the result which would be expected from acquired immunity. The latter hypothesis does not explain why the death rate should decline."
J. W. Gowen and R. G. Schott.

.[Author's comment: Even acquired immunity is genetic.]

Appendix D:
Mutagenic pollution:
A very short history

"Down syndrome, the most common genetic cause of intellectual disabilities, was first described in 1866, during an era of great change in our understanding of genetics and evolution. Because of its importance, the history of research on Down syndrome parallels the history of human genetics. In many instances, research on Down syndrome has inspired progress in human genetics. In this article, we describe the interplay between advances in the understanding of genetics and the understanding of Down syndrome from its initial description to the present, and on the basis of this historical perspective, speculate briefly about the future of research on Down syndrome."
Nature Reviews Genetics, February 2005.

In the Stone Age the invention of the stone axe, fire, and the burning of woodlands created the first man-made mutagenic pollution: smoke. Of course, the invention of the saw (4,000 BC) increased the production of mutagenic smoke.

Fired brick (4th century BC-) meant more gene-damaging smoke. The energy-providing use of natural gas and oil (4th century BC-) added to mutagenic pollution. Gas warfare (China 4th century BC-) was not good for our DNA.

Salting and smoking (techniques of food preservation) invented by the Chinese, of course, put a lot of mutagens onto our food. (Even NaCl is mutagenic.)

Gunpowder (9th century-) produced a large amount of gene-damaging smoke. War technology is one of the greatest producers of mutagenic smoke. Rockets (1241-), the gunpowder blowpipe (1304-), the gunpowder cannon (1320s-), the musket (1570-), the revolver (1630-), the modern gun (1807-), the repeater rifle (1860-), the modern war machines (20th century-), the atomic bomb (1945-), were new stages in the genetic catastrophe.

The politicians and the industry did not want to develop the atomic bomb, but the leading scientists worked hard to persuade them that the atomic bomb was important for the war. This is a good example of how science rules the World.

Cigar and cigarette smoking (Europe 1518-) enlarged the genetic catastrophe, putting several hundred mutagens into our bodies.

Public lighting (street oil lamps, 1700-) polluted our streets with mutagens and carcinogens. The widespread use of coal (1800-) certainly was a genetic tragedy. Matches ('safety' matches 1855-) directly and indirectly represented additional "progress".

Serious pollution by heavy metals (heavy metals are mutagenic) started in the Copper Age, and increased in the Bronze Age (bronze = copper + tin) and the Iron Age. Metal money (8th BC-) meant serious mutagenic heavy metal pollution. Metal pipes made out of lead (3rd century BC-) meant serious mutagenic lead pollution. The use of porcelain (1st century-) meant increased mutagenic heavy metal pollution of our bodies and our environment. Color printing (9th century-) spread more gene-damaging heavy metals in the environment (including our bodies). The invention of the mercury thermometer (1672) was a genetic catastrophe.

Canned food (1815-) and certain chemical pesticides (1800-) put more mutagenic heavy metals in our food and our bodies. Non-synthetic chemical pesticides (1800s-) such as creosote and lead arsenide were bad for our DNA. In more modern times aluminium (1850s-) directly, but mostly indirectly, was a great "improvement" of our scientific civilization. Needless to say, the general genetic catastrophe is a by-product of our scientific civilization. Production and use of superphosphates (fertilizer) (1900-) pollutes our food and bodies with mutagenic and carcinogenic cadmium.

Modern synthetic pesticides (1950-) were a direct genetic tragedy. Today more and more of these terrible man-made molecules are shown to be mutagenic and carcinogenic.

The Industrial Revolution (in the 1700s and 1800s) increased the production of numerous gene-damaging chemical substances. All new inventions caused increased industrial pollution and increased mutagenic/carcinogenic pollution. The inventions of modern technology (1850-) created new ways of destroying our genes and chromosomes. Here is a list of some of the gene-damaging inventions of the Industrial Revolution:

Asphalting roads (France 1730-) (Asphalt contains mutagenic/carcinogenic pollution.)

The steam engine (1769-)

Machine tool (1775-)

Chlorination and ozonization (1700s-)

Lead bullet (1700s-)

Metal cartridges (1799-)

Metal-coated bullet (1800s)

The locomotive (1813-)

Calculating machine (commercial) (1820-)

Public railway (1825-)

Steam turbine (commercial use, 1831-)

High-explosive shell (1837-)

Stamping (sheet-metal manufacture) (1838-)

Vulcanization of rubber (1839-)

Metal hulls and propeller propulsion (1843-)

Metal construction (1848-)

Armour-plating of warships (1850s-)

Cannon with grooved bore (1858-)

Photography (1858-)

Drilling tower (in the West) (1859-)

The London Underground (1863-)

(Despite the gene-damaging sulphurous gases emitted by the system, the London Underground was immediately successful. It transported some 10 million passengers in its first year of service in 1863. The London Underground was electrified in 1890. Note that electromagnetic radiation breaks our chromosomes.)

Transcontinental railroad (1869-)

Modern bicycle (1874-)

Standardization and scientific management (1880-)

The electrical transformer (1880s-)
Gasoline-driven automobile (1885-)
The electric generator (1890s-)
Fibreglass (1893-)

The electrical transformer and the electric generator produce electromagnetic radiation breaking our chromosomes.

The 20th century was, of course, a global genetic catastrophe. Millions of persistent man-made mutagenic chemicals are now found absolutely everywhere in the environments of the World, including our food and our bodies. We now know that the persistent organochlorines damage our DNA. The cities and the roads of the world are a genetic Hell filled with polluting cars and chimneys, and electromagnetic smog. All kinds of things that damage our genes and chromosomes. Probably as much of 90 percent of known inventions causing mutagenic pollution (including mutagenic radiation) date from the 20th century. Below is a list of some of the gene-damaging (directly or indirectly) inventions of the 20th century:

Gas turbines (1900-)
Low-cost cars (1908-)
The New York metro (1912-)
Electrification of the railways (20th century-)
Modern ships (20th century-)
Modern submarines (20th century-)
(The first military use of a submarine dates back to 1775, to the American War of Independence.)
Synthetic glues (1930-)

(The old adhesives were non-mutagenic and based on plaster, resin, starch, fish glue, casein gelatine, wax, eggwhite, etc.)
Nuclear energy (electric power) (1951-)
(In the "Radiation Area" of our nuclear industry the workers' chromosomes are being damaged. Hundreds of man-made radioactive isotopes have polluted the whole World. Science explains away this fundamental genetic catastrophe.)
Modern medicines (20th century-)
(Many modern chemical medicines are known to be gene-damaging. Pharmacopoea Danica 1933 tells us that creosote and mercury are good medicines! Today diagnostic use of X-rays damages our chromosomes. Never ever trust science!)
The compact cassette (1962-), the modern computer (1968-), the color home video recorder (1972-), etc, etc meant enormous mutagenic/carcinogenic pollution. A burning PC, say, produces extremely damaging pollution. Even a non-burning PC and a non-burning TV, for examples, emit mutagenic/carcinogenic chemicals.

Directly and indirectly, the products and by-products of the Electric Age (1879 – 1946), the Electronic Age (1947 – 1972), and the Information Age (1973-) represent a general genetic catastrophe of the World.

In 1962 the most important book of the 20th century was published: *Silent Spring* by Rachel Carson. Science had created and produced enormous amounts of extremely damaging pesticides. Carson made the world aware of this global catastrophe. She was persecuted because science told us that she was wrong. But the environmentalists and the politicians reacted to a certain extent. In 1963, the first

"Clean Air Act" was passed to control air pollution, and in the 1960s environmentalism became a political movement. However, the general genetic catastrophe of the world has exploded since the 1960s. Science answered Carson by inventing and producing an enormous number of new dangerous and persistent pesticides and other kinds of mutagenic and carcinogenic environmental poisons.

The 1972 book *A Blueprint for Survival*, written by a large number of scientists, claimed that "social disintegration is a major cause of mental disease." Most other claims in this well known book are wrong as well, history has shown us, and this book systematically avoided the issue of gene/genetic damage. Unbelievable but true: Science did – and still does – systematically neglect, avoid, boycott, misunderstand, explain away, or cover up the fundamental problem of general DNA-damaging pollution and general gene/genetic damage.

The oceans are an ultimate accumulation site of the chemical and radioactive products and by-products of our scientific civilization. Even the artic is now totally polluted by mutagenic and carcinogenic chemicals. The polar bears contain more mutagenic environmental poison than many types of dangerous industrial waste. Arctic seagulls filled with mutagenic pollution are falling dead to the ground. Also the arctic fishes are poisoned by mutagenic chemicals. Science neglects this fact. Never trust science at the dining table.

The effects of mutagens are not separable, but instead additive or mutually reinforcing. Rapid cumulative genetic collapse of our scientific civilization is now a real threat. But science is silent. Politicians, labor leaders, industriy leaders, and science fight together for the creation of new polluting industry based on new science-based technology and new science-created mutagenic chemicals. Their decisions are always scientific, and their advisers, consultants, and experts are the scientists. No one listen to "unscientific" people. Remember that scientific management was established in the 1880s, and note that fact is: the general genetic catastrophe exploded in the 1880s.

As a matter of fact, basically, science has created the general genetic catastrophe. How? Partly by replacing non-mutagenic natural products by mutagenic synthetic products. Examples: Plastics replaced wood products. Detergents replaced soaps made from natural fats. Synthetic fibres replaced natural fibres. Chemical fertilizers replaced organic manure. Fossil fuels replaced wood. Nuclear energy replaced fossil fuels. Etc. Etc. Nature does not produce a single molecule for which there is not an enzyme capable of breaking it down, in order to perpetuate the cycle of life, growth, death, and decay. Synthetic products cannot normally be broken down in this way. Therefore, even the slightest amount of synthetic products constitutes serious pollution, because these chemicals find their way into our bodies, where they attack our genes, chromosomes, cells, and organs. The activities of scientists and industrial man are having a very serious effect on our bodies and our society. "Even in the realms of science, take nothing at face value." (John Waller: *Fabulous Science*.) "The myth of scientific detachment." (John Waller: *Fabulous science*.)

Appendix E:
DNA protection: The very basics

"One puff of tobacco smoke can damage DNA."
Researchers at University of Pittsburgh.

Each human cell sustains some 10,000 DNA mutations every day. If it were not for DNA repair enzymes and other repair mechanisms, these mutations would lead to catastrophic accumulation of DNA damage. But there is a limit to the cell's ability to repair DNA alterations. Therefore, it's important to protect genes against mutations. The most effective way to protect against mutagenic activity is to live in an environment that does not contain manmade mutagenic pollution (including manmade mutagenic radiation). However, cells lacking micronutrients (vitamins, minerals, etc.) may suffer increased DNA damage, because micronutrients are involved in DNA maintenance and repair. And DNA instability caused by micronutrient deficiencies increases DNA's sensitivity to mutagens. The most effective nutrient to protect against gene mutation is chlorophyllin. One study showed that chlorophyllin suppressed mutagenic activity by more than 90 percent. (So eat much green leafy vegetables, for example.) Curcumin (an antioxidant, powder) protects the cells against the entry of chemicals that mimic like oestrogen, for example: pesticides, DDT, and dioxin. Curcumin can kill cancer cells.

Food (plant) fibers cause rapid excretion and help to minimize exposure to manmade mutagenic chemical pollution. Also: Fibers bind some mutagens and prevent them from being absorbed by our cells.

Free radicals are unstable molecules that can steal particles from other molecules to make themselves more stable. Free radicals are produced by normal metabolism. But chemical pollution, radiation, heavy metals, smoking, alcohol, etc. can dangerously increase the production of free radicals. Antioxidants counteract this process by binding the free radicals, transforming them into non-damaging molecules, and repairing damaged cells. However, both lack and excess of antioxidants can increase the problem of gene damage. And too much antioxidants can decrease the effectiveness of the immune system. (The immune system can create free radicals on purpose to neutralize viruses and bacteria.)

Antioxidants in green plants, colored fruits, and colored vegetables may protect against gene damage by removing mutagenic free radicals. Fruits (fresh is best) like cherries, red grapes, blueberries, strawberries, kiwi, oranges, red grapes, papaya, pomegranate, avocados, etc. are known to reduce DNA damage. Vegetables like broccoli, watercress, brussels sprouts, cabbage, cauliflower, garlic, red onions, yellow onions, asparagus, green beans, black beans, beets, carrot, tomato, etc. are also known to reduce DNA damage. Herbal indredients such as green tea, reveratrol, curcumin, ashwagandha, etc. are known to reduce DNA damage, as well. Also dark chocolate, coffee, olive oil, sardines, and yogurt do contain antioxidants and so may protect against

DNA damage. Vitamin A, betacarotene, Vitamin C, vitamin E, zinc, selenium, green tea extract, grape seed extract, and ginkgo biloba are well known antioxidants.

When scientists discovered that nitric oxide (NO) could be produced in the body, they thought that this compound was responsible for causing damage to DNA. However, it has since been learned that when NO reacts with another free radical, the superoxide anion, something more dangerous is formed: peroxynitrite. Peroxynitrite interacts with DNA, causing DNA damage, and may therefore cause cells to die by apoptosis. (Apoptosis is a normal processes, but it may not be desirable when triggered by peroxynitrite.) The consequences of damage from peroxynitrite are a variety of conditions: cancer, neurodegenerative disorders, stroke, heart failure, diabetes, Alzheimer's disease, fibromyalgia, and PTSD. Ecklonia Cava Extract (ECE) is effective against peroxynitrite.

Avoid fried-food. Avoid food preserved in nitrates and hydrogenated oil. Use temperatures lower than 180 degrees Celsius to prevent brownness (to prevent the creation of PAH). Avoid trans fats. Fumes in a modern kitchen are, in general, mutagenic. In addition, the extremely powerful (and alternating) electromagnetic radiation from modern electric cookers and ovens, for example, is mutagenic. So, avoid mutagenic smoke and mutagenic fumes. And avoid mutagenic electromagnetic radiation. Cooks, high voltage workers, and radar workers do have damaged DNA, for example chromosome damage. Researchers found that the rate of suicide among 5,000 electricity utility workers who were exposed to low frequency electromagnetic radiation was double that of a control group of 5,000 people. The effect was particularly noticeable among young workers. (Journal of Occupational and Environmental Medicine, March 15, 2000.)

Important links to information on food mutagens:
www.llnl.gov/str/Food_Mutagens.intro.html
www.copperwiki.org/index.php?title=Mutagens

Synthetic materials are often mutagenic. Therefore use natural materials and iron (not stainless steel, for example, because heavy metals are mutagenic.) Natural materials are stone, flint, sand, clay, soil, wood, bark, leather, wool, silk, cotton, hemp, jute, etc.

Use ceramic cookware, tableware and kitchenware that contain no lead, cadmium and other heavy metals. Ceramic ware is made of ceramic clay and natural materials. By the way: Porcelain (chinaware) is a kind of ceramic. Use only lead free glass (sodium glass or potash glass). Remember that fumes from heated teflon are mutagenic and carcinogenic.

A modern bathroom and a modern kitchen are chemical hells. But Sunlight soap is pretty natural, produced and sold since 1884, and can be used for everything: bath, body wash, hair wash, laundry, dishes, house cleaning, etc. Sunlight soap and the washing machine: Sunlight soap, grated and boiled in water, left to form into a gel, and used in the front loader of the washing machine.

Chlorine and its by-products are mutagenic and carcinogenic. Natural organic substances, such as, for example, vegetables and fruits, can form cancer causing chemicals when combined with chlorinated water. Do not drink chlorinated water, and do not wash yourself in chlorinated water. A warm shower releases the chlorine in vapor form for you to inhale.

Natural homemade stain removers are extremely effective. There are many Natural Stain Removal Guides on the Internet. How to get oil-based paint off your skin? Rub your hands (skin) with butter or margarine. How to clean a car mechanic's hands? Again, use butter or margarine. (You may add a little sugar to get a scrubbing effect.)

In 1985 the Environmental Protection Agency (EPA) reported that toxic chemicals found in the air of almost every American home are three times more likely to cause some type of cancer than outdoor air pollutants.

Using natural building materials you can build a home with non-mutagenic, renewable and locally available materials. Our modern homes and workplaces are mutagenic. When there are much mutagens around, use protective mask, protective gloves, and protective clothing. This is all the more important now when the human population is subjected to a catastrophic humanmade mutagenic pressure due to medical diagnostics, industrial accidents, occupation, and growth in environmental contamination.

Modern roads and modern streets are mutagenic hells. Gasoline-fueled vehicle exhaust have been shown to directly damage DNA. Also, diesel exhaust particles have been shown to directly damage DNA. Stay away from vehicle exhaust. Stay away from smoke from all kinds of fires. Stay away from torchlight processions; the smoke from a flaming torch is mutagenic. Tear gas (CS gas) is mutagenic, as well, so avoid violent demonstrations. (The agent CS can alkylate sulfhydryl groups and, possibly, DNA. As such, it is potentially mutagenic. This agent has not, however, been well studied for its genetic effects. Some researchers have shown CS to be mutagenic in the Ames test.)

Hold your breath when you pass through a cloud of black smoke from car exhaust. You should never inhale the smoke from any burning substance. If possible, walk on the side of the road where the wind is blowing from, to avoid as much exhaust as possible. Never run or jog along roads where cars are driving. Remember to wear bright and reflective clothing, and walk "defensively."

UVB and UVC light can cause direct DNA damage. (Arc welding, for example, produces UVB radiation.). Be careful.

Today gasoline contains ca. 1 percent benzene. Benzene is mutagenic and carcinogenic. And benzene causes chromosome breaks. The American Petroleum Institute (API) stated in 1948 that "it is generally considered that the only absolutely safe concentration for benzene is zero." Nevertheless, today, the whole world population is inhaling mutagenic benzene every day. Benzene is now used to make drugs, pesticides, etc. In the 1800s and early 1900s benzene was used as an after-shave lotion. Gasoline often contained several percent benzene before lead replaced it as an antiknock additive around 1960. Many students were exposed to benzene in schools and universities while working in laboratories with little or no ventilation. Before ca. 1980 benzene was widely used as solvents, spot removers, and as a component (or contaminant) in many consumer products. In the US some 100,000 sites have benzene soil and groundwater contamination. In China (2005) benzene contaminated a large river. In the UK (2006) they found benzene in soft drinks. Benzene *alone* represents a global genetic catastrophe. Stay away from benzene contaminated products and sites.

PVC (polyvinyl chloride) commonly known as vinyl, is a plastic used in many products: Toys, teething rings, window shades, medical supplies, water pipes, wraps, siding, flooring, rain gear, shower curtains, etc. PVC contains phthalates, lead, cadmium, light stabilizers, heat stabilizers, anti-oxidants, barium, and other chemicals, and burning PVC produces dioxins. Therefore PVC *alone* is causing a global general genetic catastrophe. So: Avoid PVC products. Avoid vinyl products. Ban PVC toys. Stop manufacturing PVC toys. Stop selling and buying PVC toys. Discourage children from chewing plastic. Do not microwave food in plastic containers or plastic wraps. Dentist must not use coating treatment containing bisphenol A. Eat lower on the food chain.

The Fossil Fuel Age is soon history. The Fossil Fuel Age represents, in many ways, a global general genetic catastrophe. The whole world should as soon as possible totally change to clean and renewable energy. Every high school student knows all the important types of alternative energy: Solar energy, wind energy, hydroenergy, classical tidal energy, underwater tidal/ocean energy, geothermal energy, bioenergy (energy forests, methane from farms and organic waste), salinity gradient energy, hygroelectricity (pulling electric charges out of humid air), heat pump systems (energy savers). And let us not forget animal muscle energy and human muscle energy.

What about nuclear fusion energy? Well, nuclear fusion energy produces huge amounts of radioactive pollution and other kinds of pollution, contrary to what most people believe. Moreover, the production of fuel (deuterium and lithium) is too costly and too polluting, contrary to what most people believe. But scientists and researchers want money, work, and challenge. This book shows that the Atomic Age is a global general genetic catastrophe. Sweden, for example, has decided to end their part of the Atomic Age.

If we change the world economy to short-travelled food, short-travelled goods, and short-travelled raw materials, etc., then it's not necessary to use fossil fuel driven ships. Sailing ships (large wind-powered vessels) can do the whole job using local wind and global air and sea currents. Even today, technically, air traffic can be driven using non-fossil fuel. Hydrogen fuel cell cars, and even hydrogen fuel cell ships, can be built today. We must go to work now. Why? Simply because of the existence of the general genetic catastrophe and the existence of the general climate catastrophe!

Build eco-friendly houses! Practice organic farming! (See articles and videos on the internet.) Yes, we can! Back to

Voluntary Simple Living (Live simple. Live free. A debt free life. Save money by doing it yourself. Voluntary simplicity is about freedom. It's about owning your own life.) Back to Frugal Living (Frugality is living with less of what money can buy.) Yes, we can! *Intermediate technology* (the term was coined by E. F. Schumacher) will be extremely important in the future.

Finally, some good alternative literature:
Back to Basics: A Complete Guide to Traditional Skills (Third Edition, newly updated) by A. R. Gehring. *The Ecyclopedia of Country Living* (Ninth Edition, newly updated) by C. Emery. *Country Wisdom & Know-How* by The Editors of Storey Publishing's Country Wisdom Boards. *Primitive Skills and Crafts: An Outdoorsman's Guide to Shelters, Tools, Weapons, Tracking, Survival, and More* by R. Jamison and L. Jamison. *The Modern Hunter-Gatherer: A Practical Guide to Living Off the Land* by Tony Nester and J. Cole. *SAS Survival Handbook for Any Climate and Any Situation* by J. Wiseman. *Alternative Agriculture* by Committee on the Role of Alternative Farming Methods in Modern Production Agriculture, and National Research Council. *Alternative Energy Demystified* by S. Gibilisco. *Doable Renewables: 16 Alternative Energy Projects for Young Scientists* by M. Rigsby. *Wind Power for Dummies* by I. Woofender. *Solar Power Your Home for Dummies* by R. DeGunther. *Energy Efficient Homes for Dummies* by R. DeGunther. *Toward a Zero Energy Home: A Complete Guide to Energy Self-Sufficiency at Home* by D. Johnston. *The Crash Course: The Unsustainable Future of Our Economy, Energy, and Environment* by C. Martenson.

Appendix F:
Socialist and non-socialist eugenics

"Socialists, with their belief in planning and their readiness to put the state in the position of power over the individual, were ready-made for the eugenic message [from mainstream science]."

Matt Ridley

"John J. Donohue III (professor of law at Stanford University) and Steven D. Levitt (professor of law at University of Chicago) say: "Legalized abortion contributed significantly to recent crime reductions." Their study was published in the second quarter, 2001 *Harvard's Quarterly Journal of Economics*. According to the authors, "Crime began to fall roughly 18 years after abortion legalization." Since 1991, homicide rates are down 40 percent, violent crime and property crime are down 30 percent. After controlling for other factors, Donohue and Levitt conclude: "Legalized abortion appears to account for as much as 50 percent of the recent drop in crime." Rich Deem (See: www.godandscience.org)

"Why were Roma forcibly sterilised in Norway up until the seventies?" Question by Halskiisaklink, on April 14th, 2008 (See answer later.)

Evolution of the human species is nature-made "breeding" of humans. Eugenics is man-made breeding ("evolution") of humans. Breeding of animals (domestic animals) have gene-damaged the bred animals. A couple of examples: A domestic pig is a severely gene-damaged "version" of genuinely wild pigs. A dog is a severely gene-damaged "version" of wolves. Why? Because breeding only breeds very few of the tens of thousands characteristics of an organism, while evolution protects and evolves *all* characteristics of an organism. Eugenics is breeding and may therefore, in the long run at least, be damaging to the human gene pool. Moreover, eugenics cannot remove mutagens from the human body, and today manmade mutagens and manmade gene damage are the overwhelming problems, destroying our DNA on a daily basis. Regarding genetic diseases causing early death and genetic diseases causing having no children, eugenics is almost superfluous. More than 50 percent of very young human fetuses are automatically and naturally aborted due to chromosome and DNA damage. Moreover, severely gene-damaged eggs and sperms are aborted, and we have a lot of stillbirths and deadly too-early births. Not to forget: The so called "anti-eugenicists" were the most eager fighters for free abortion. Now we do have free abortion, and several hundred million abortions have taken place in the world during the last 50 years, killing much more bad genes than good genes, we now know from several studies. This amount of eugenics is many times bigger than all earlier eugenics projects combined. Today we can hunt for gene-damage in potential parents, and ask them not to add pain and suffering to an already painful and suffering world. But this is the *opposite* of eugenics, because *eugenics is breeding the elite.* Here we are, firstly, not breeding at all. We are only fighting genetic destruction. And that is not breeding (eugenics). Secondly, we are not helping an elite. We are helping a non-elite. As we know, "lower classes" are more gene-damaged than "higher classes." And, as we also do know, genetically unhealthy people are earning less money than genetically healthier people. Hopefully I now have clearly explained why this book is not about eugenics. But people, in general, do not know what eugenics is all about. Therefore I've included an appendix on eugenics in this work.

Since C. Darwin and until a few years after WWII mainstream scientists were perpetually urging eugenic actions by the governments. Darwin's first cousin F. Galton coined the term *eugenic* in 1885. "Let us breed from the best and not from the worst specimens of humanity," said Galton. But even before Galton, in 1857 the French physician, B. A. Morel noted that, due to improvements in public health many infants who would have previously died were now surviving. He concluded that this would result in a deterioration in traits such as physique, intellect and moral character. Darwin's son, Leonard, was president of the Eugenics Society. Karl Pearson, a socialist and statistician, was the first and most influential follower of Galton. By 1900 eugenics was extremely popular in Europe. The socialists said that it was not the individual that must be eugenic; it was the nation. They said the state must have a say in who should breed or who should not. (In 1946 in Norway, with a socialist government, my mother and my father had to lie about genetic diseases in their families in order to be allowed to marry.). In general, the non-socialist eugenics meant that each individual strove to find a mate with a good mind and a good body. But the socialists won the battle. And in Germany most scientists joined the National Socialist Party. Norway, Sweden, Finland, Iceland, and Estonia sterilised an enormous number of people. The Nazi-friendly Sweden alone sterilised almost as many people as the US. Sweden and Germany implemented compulsory sterilisation laws in 1934. Germany sterilised some 400,000; relatively fewer than the Nordic countries. However, the National Socialists and the socialists (communists) in the Soviet Union mass-killed gene-damaged people by the millions. The Nordic and the National Socialist sterilisation laws were praised all over the world. The socialists' eugenics leader in the UK, Karl Pearson, said: "What is social is right, and there is no definition of right beyond that." In the US, the famous socialist and scientist H. J. Muller wanted to see children carefully bred with the character of Marx and Lenin.

But suddenly, after WWII, both socialists and non-socialists, and many mainstream scientists, were against eugenics. Why? Well, the non-socialists, mainly in the US, had beaten the super-eugenic Germany (and the racist Japan), and now Germans who had killed and sterilised people in the name of eugenics, had to be punished by law. Moreover, non-socialists saw eugenics as an infringement of personal liberty. The Catholic Church was also influential. In countries where the influence of the Catholic Church was strong, there were no eugenic laws. In the UK, North America, and Sweden the Jews had survived. They had a story to tell, and they were rich, intelligent, and influential. What about the socialists? Well, the socialists now discovered that there were much more feeble-minded people among the working class than among other classes of society, and they did not want to lose votes. The socialists suddenly saw eugenics as a form of war on the working class. Just as the non-socialists, the socialists were not against eugenics itself. And even just after WWII eugenics was still mainstream science. However, now the leftists believed (wrongly) that human nature was explained by environmental factors, not by heredity and biology. Undoubtedly, it was Hitler, Mengele, and Holocaust that killed the eugenics in Europe and North America. Eugenics was no longer PC (Politically Correct).

But soon liberals, socialists and communists were on the road again, both in Europe, North America, and China. The New Eugenics was Free Abortion (in China compulsory abortion). Since WWII abortions of human fetuses in the US alone, have killed more than 50 million humans (fetuses), mostly from classes with relatively much violence and criminality. And the new eugenics worked extremely well. Because violent and criminal behavior is largely genetic, the new eugenics caused violence and criminality to drop more than anyone had expected. Today the New Eugenics is much more than killing fetuses. Even modern genetics is simply a part of the New Eugenics, but modern genetics may also be dysgenic. Medicalized reproduction may be both eugenic and dysgenic. Prenatal testing (PT) is eugenic in that its aim is to reduce the number of people with genetic disorders. Preimplantation genetic diagnosis (PID) is an eugenic improvement upon PT because multiple embryos are available. Human biotechnology is both eugenic and dysgenic. (The society as a whole is dysgenic.) The state newborn genetic screening, the state genetic registries, the state comprehensive genome screening of newborns at birth, the new state intensification of the medical surveillance, etc., show that the good old socialist and liberal Pearsonian eugenics is back on track. Even in Nazi Germany abortion by fit women was illegal. But they who still love dysgenics more than eugenics can find comfort in the fact that the societies and the cultures of the world are still catastrophically dysgenic. I'm, of course, taking about the manmade global general genetic catastrophe due to manmade local and global mutagenic pollution. Socialist China is fighting hard against the production and spreading of genetic diseases, as always. The minister of public health in China said that births of inferior quality are serious among the revolutionary base. In China, The Maternal and Infant Health Care Law makes premarital check-ups compulsory, and if necessary, abortion compulsory. Echoing the European influential socialist and leading scientist Karl Pearson: "The Chinese culture is quite different, and things are focused on the good of society, not the good of the individual." (Xin Mao.)

Back to Norwegian eugenics:

Copy from Gaydarnation.com, News, December 12, 2000:

Quote:

"Norway Sterilised Gays for 35 Years.

Norway castrated and sterilised 400 people, including many gay men, between 1934 and 1969, according to new research.

Per Haave, a historian at Oslo University, had access to Norway's public health records and found that the Norwegian authorities sterilised 414 people over a period of 35 years. He said that sterilisation was seen as a way of treating rapists, mental patients, people suffering from epilepsy, and homosexuals.

Under the policy, 370 men had their testicles removed and 44 women had their ovaries taken away. Many were aged under 20. He said that the sterilisations peaked in the late 1940s and that more than 100 people were castrated between 1948 and 1950.

Haave said: "Castration was to prevent sexual crimes, but it quickly took on a much broader scope." He explained that sterilisation, "was perhaps used most of all on psychiatric patients, on those in mental institutions and in homes for young criminals. Among the men there were many homosexuals".

Haave's research follows revelations in 1997 that Sweden had sterilised 63,000 people between 1935 and 1975 in a eugenics experiment. Many had been sterilised against their will."

Unquote.

Copy from www.answerbag.co.uk:

Quote:

"Question by Halskiisaklink, on April 14th , 2008:

"Why were Roma forcibly sterilised in Norway up until the seventies?"

Answer by Tucson, on on April 14th , 2008:

"The Roma have been discriminated against and treated horribly for centuries upon centuries. It was the gypsies that were blame for many to all of a village's woes, it was the gypsies who cohorted with the Devil, and it was the gypsies who stole and ate children. These past prejudices have carried forth into modern times, even within months or days to the present day.

The sterilization movement really got started with the Nazis and their love and belief in Eugenics. Eugenics is a social philosophy which advocates the improvement of human hereditary traits through various forms of intervention, basically breeding a better human.

Part of the Nazi regime's ethnic cleansing of the undesired masses, the Roma included, was their practice of forced sterilisation. In recent years, Romani women in Bulgaria, the Czech Republic, Hungary, Romania and Slovakia have also been subjected to forced sterilisation. From the 1970s until 1990 the communist government of Czechoslovakia sterilised Romani women as part of an official policy to reduce their 'high, unhealthy' birth rate.

They implemented their policy through programmes that provided monetary incentives for women to undergo the operation, and condoned misinformation and coercion. Although it has been assumed that the practice ended in the 1990s, the European Roma Rights Centre says that there is evidence that coercive sterilisations continue to date.

As horrible as this may seem, forced sterilisations have been enacted upon the mentally disabled, the blind, gays and lesbians, ethnic minorities, religious minorities, and anyone else who wasn't rich/white/powerful.

The final question, why? Genocide is illegal so preventing the children from existing at all is the next best step."

Unquote.

Finally I wish to mention a couple of *non-eugenic* success stories:

The number of children born with thalassemia in Cyprus is now virtually zero because Cyprus acted on a recommendation from WHO in 1973 (compulsory carrier screening and counceling). The Orthodox Church of Cyprus actively supported WHO, the Government, and the people of Cyprus.

Cystic fibrosis has been almost eliminated from the Jewish population in the US thanks to the Committee for Prevention of Jewish Genetic Disease which organizes the testing of schoolchildren's blood, and later advices against marriage if both persons are carrier of the same gene damage. Unbelievably but true: This voluntary *non-eugenic* policy was criticized in 1993 (sic) by the leftist-dominated New York Times, as eugenic.

Appendix G:
Quotes that can help us stand up

"The white man [...] is a stranger who comes in the night and takes from the land whatever he needs. The earth is not his brother but his enemy. [...] Continue to contaminate your bed, and you will one night suffocate in your own waste."
(Chief Seattle of the Duwanish tribe of American Indians written in1855 in a letter to the president of the USA.)

"The one process now going on [...] is the loss of genetic and species diversity by the destruction of natural habitats. This is the folly our descendants are least likely to forgive us."
E. O. Wilson

"One of the greatest diseases is to be nobody to anybody."
Mother Teresa

"Don't let anyone steal your dream. It's your dream, not theirs."
Dan Zandra

"If we all did the things we are capable of doing, we would literally astound ourselves."
Thomas Alva Edison

"If you really put a small value upon yourself, rest assured that the world will not raise your price."
Author unknown

"Many great ideas have been lost because the people who had them could not stand being laughed at."
Author unknown

"The fear of being laughed at makes cowards of us all."
Mignon McLaughlin

"We probably wouldn't worry about what people think of us if we could know how seldom they do."
Olin Miller

"Other people's opinion of you does not have to become your reality."
Les Brown

"Courage is the power to let go of the familiar."
Raymond Lindquist

"If one is forever cautious, can one remain a human being?"
Aleksander Solzhenitsyn

"Nothing diminishes anxiety faster than action."
Walter Anderson

"The way you overcome shyness is to be so wrapped up in something that you forget to be afraid."
Lady Bird Johnson

"Nothing so much prevents our being natural as the desire to seem so."
Francois

"Nobody realizes that some people expend tremendous energy merely to be normal."
Albert Camus

"Nobody can make you feel inferior without your consent."
Eleanor Roosevelt

"If you doubt yourself, then indeed you stand on shaky ground."
Henrik Ibsen

"Men are not against you; they are merely for themselves."
Gene Fouler

"There are offences given and offences not given but taken."
Izaak Walton

"Let me never fall into the vulgar mistake of dreaming that I am persecuted whenever I am contradicted."
Ralph Waldo Emerson

"I don't have a lot of respect for talent. Talent is genetic. It's what you do with it that counts."
Martin Ritt

"We are driven by five genetic needs: survival, love and belonging, power, freedom, and fun."
William Glasser

"Science is an edged tool, with which men play like children, and cut their own fingers."
Arthur Eddington

"Not to be absolutely certain is, I think, one of the essential things in rationality."
Bertrand Russell

"The problem with simple arguments is that they may be difficult to explain."
Karin Erdman

"I have hardly ever known a mathematician who was able to reason."
Stephen Hawking

"If at first an idea isn't absurd, there's no hope for it."
Albert Einstein

"Now, my own suspicion is that the universe is not only queerer than we suppose, but queerer than we *can* suppose."
J. B. S. Haldane

"We do not know why they [elementary particles] have the masses they do; we do not know why they transform into another the way they do; we do not know anything!"
George Gamow

"What is it that breathes fire into the equations and makes a universe for them to describe .... Why does the universe go to all the bother of existing?"
Stephen Hawking

"Somehow, there must be, wondered Einstein, a kind of 'pre-established harmony' between human inventive conceptual imagination and aspects of reality itself."
Arne Naess

"The most beautiful thing we can experience is the mysterious. It is the source of all true art and science."
Albert Einstein

"Science without religion is lame, religion without science is blind."
Albert Einstein

Appendix H:
A list of websites

Found on the World Wide Web:

"I think it's morally irresponsible and cruel to knowingly conceive a child when you have a genetic disorder that your children would be likely to get."

"Many people with genetic disorders choose not to have kids anyway."

"Adoption is always a choice."

"People should stop having so many kids. The world is over populated."

"Question: Estimate the number of people with genetic diseases in the world. Answer: 6 billion. Almost everyone has or will have a genetic disease in their lifetime."

The websites below are shocking. They range from reporting tobacco snuff causing gene damage, to chromosome damage in oil spill workers, and from scientists question safety of new airport scanners, to chromosome damage in liver cells due to low dose radiation. They range from frequent flying causing chromosome damage, to chromosome damage from non-ionic contrast media. And so on, and so forth.

http://en.wikipedia.org/wiki/List_of_geneticists (LIST OF GENETICISTS)
www.medicinenet.com/genetic_disease/article.htm
www.buzzle.com/articles/genetic-disorders-in-human.html
www.chej.org
www.ejnet.org

http://rarediseases.about.com
http://rarediseases.info.nih.gov
www.nlm.gov/medlineplus.rarediseases.html
www.rarediseases.org
www.hon.ch/HONselect/RareDiseases
www.eurordis.org
www.orpha.net
www.rarediseaseday.org
www.rare-disease.eu/2010/index.php
www.raredisease.org.uk
www.socialstyrelsen.se/rarediseases
www.eucerd.eu
http://webhealth.com/wiki/Chromosomal_Abnormalities
www.rarechromo.org/html/home.asp
www.niehs.nih.gov
www.ems-us.org
www.eems-eu.org
www.ukems.org
www.iaems.net
www.j-ems.org
www.gems-nc.org
http://www.ejnet.org/dioxin/
http://cancerres.aacrjournals.org/content/38/3/608.full.pdf
www.irb.hr/nato-savariver/files/abstracts.pdf
www.philstar.com/Article.aspx?articleId=636753
&publicationSubCategoryId=80
www.cirs-reach.com/Toys/index.html
http://product-testing.eurofins.com/media/17636/
Toys%20Safety%20-%20en.pdf
www.llnl.gov/str/Food_Mutagens.intro.html
http://calendar.publishing.uwa.edu.au/latest/parte/constitutions/
other/universitysafety/termsofref/carcinogeniccon
www.physorg.com/tags/mutagenic+substances/
www.springerlink.com/content/3568736003363382/

http://wmr.sagepub.com/content/6/1/149.abstract
www.thaindian.com/newsportal/tag/
    mutagenic-substances
www.cytochrome.net/showabstract.php?pmid=3547110
www.kemi.se/templates/Page____4018.aspx
www.13.waisays.com/cooking.htm
www.radford.edu/fpc/Safety/HazCom/chp5.htm
http://answers.yahoo.com/question/index?qid=
    20081106160220AAGTSqL
http://osh.sm.ee/legislation/cancirogens.htm
http://onlinelibrary.wiley.com/doi/10.1111/j.1749-6632.
    1976.tb35133.x/abstract
www.ncbi.nlm.nih.gov/pubmed/17287601
www.ncbi.nlm.nih.gov/omim/
http://pubs.acs.org/doi/abs/10.1021/es00144a005
www.unepmap.org/index.php?module=library&
    mode=pub&action=results&_stype=3&s_category=
    &s_descriptors=Mutagenic%20substances
http://rivm.openrepository.com/rivm/bitstream/
    10029/8883/1/320010001.pdf
http://osha.europa.eu/en/publications/reports/548OELs
http://mutagenicsubstance.blogspot.com/
http://ukpmc.ac.uk/abstract/MED/7003377;jsessionid=
    E80759E9339DA836BCEB8DD3C637B6EE.jvm4
www.springerlink.com/content/vvw06rm6726t8888/
www.cancer-info-guide.com/articles/
    cancer-foods.html
http://papers.sae.org/821244
www.alttox.org/ttrc/toxicity-tests/genotoxicity/
http://mutage.oxfordjournals.org/content/18/2/113.full
www.informaworld.com/smpp/content~db=all~content
    =a916052675~tab=content
www.lycaeum.org/~sputnik/Drugs/LSD/LSD.genes.html
www.rense.com/general66/sitit.htm

www.nih.gov/news/pr/feb2005/niehs-15.htm
http://esciencenews.com/dictionary/chromosome.damage
www.sciencedaily.com/releases/
  2009/06/090601092139.htm
www.trdrp.org/research/PageGrant.asp?grant_id=1614
www.google.com/hostednews/afp/article/
  ALeqM5i6uiZt-Hg9QVfmM0Ll0xm9vu-ouQ
www.nytimes.com/1995/10/25/us/chromosome
  -damage-in-the-lab-is-tied-to-a-chromium-supplement.html
http://ajrccm.atsjournals.org/cgi/content/short/176/5/505
http://agentorangezone.blogspot.com/2010/10/dioxin-
  and-chromosome-damage.html
www.restoreunity.org/chromosome_heroin.htm
www.sciencemag.org/content/190/4219/1090.abstract
www.npr.org/templates/story/story.php?storyId=126833083
www.scienceagogo.com/news/20091018230147data_trunc_sys.shtml
www.ncbi.nlm.nih.gov/pubmed/19296340
http://ic.ucsc.edu/~saltikov/bio119/readings/2003_Wegrzyn.pdf
www.pnas.org/content/94/16/8380.long
http://en.wikipedia.org/wiki/Eugenics#Modern_eugenics.2C_
  genetic_engineering.2C_and_ethical_re-evaluation
https://groups.google.com/group/dna_protection?hl=no
https://groups.google.com/group/derailed-evolution?hl=no
http://tech.groups.yahoo.com/group/EP_group/
http://health.groups.yahoo.com/group/dna_
  protection/?yguid=146684972
http://health.groups.yahoo.com/group/GGC_
  group/?yguid=146684972
www.emwatch.com
www.rationalfuturist.com/writings/fusion.html
www.gotmercury.org
www.savetherainforest.org
www.treehugger.com
http://forum.baby-gaga.com/about256936.html

www.disabled-world.com/artman/publish/article_0060.shtml

www.proquestk12.com/productinfo/pdfs/Georgia%20SBLA_
science_9-12_geneticdisorders.pdf

http://ghr.nlm.nih.gov/handbook/mutationsanddisorders/
statistics

http://en.wikipedia.org/wiki/List_of_human_genes

http://en.wikipedia.org/wiki/List_of_genetic_disorders

www.ornl.gov/sci/techresources/Human_Genome/medicine/
assist.shtml

http://kidshealth.org/teen/your_body/health_basics/genes_
genetic_disorders.html

http://ec.europa.eu/health/ph_threats/non_com/docs/overview_
national_en.pdf

www.crdnetwork.org/blog/

http://rarediseasesupport.org/

www.inforesearchlab.com/smokingdeaths.chtml

http://www.toqonline.com/blog/the-fall-of-man/

http://www.ecomall.com/greenshopping/ftoys.htm

http://anthropologyworks.com/?p=2395

http://bioinformatics.weizmann.ac.il/cards

www.symptoms101.com/med/archives/genetic_disorders/

http://www.disabled-world.com/health/pediatric/
sotos-syndrome.php

https://www.llnl.gov/str/Food_Mutagens.intro.html

Bibliography (books)

Many of the books below are shocking. Here are three examples:

Deadly Glow: The Radium Dial Worker Tragedy.

The Woman Who Knew Too Much: Alice Stewart and the Secrets of Radiation.

Slow Death by Rubber Duck: The Secret Danger of Everyday Things.

Aitken, K. J. (2010). **An A – Z of Genetic Factors in Autism: Handbook for Professionals**. Jessica Kingsley Publishers.

Anderson, M., and A. Anderson (2010). **Reagan's Secret War: The Untold Story of His Fight to Save the World from Nuclear Disaster**. Three Rivers Press.

Ashby, M. F. (2009). **Materials and the Environment: Eco-informed Material Choice**. Butterworth-Heinemann

Ayers, H., et al. (1998). **An Appalachian Tragedy: Air Pollution and Tree Death in the Eastern Forests of North America**. Sierra Club Books for Children.

Baker, N. (2009). **The Body Toxic: How the Hazardous Chemistry of Everyday Things Threatens Our Health and Well-being**. North Point Press.

Becker, J. B., et al. (1992). **Behavioral endocrinology.** MIT Press, Cambridge.

Berg, K. (1980). **Genetic Damage in Man Caused by Environmental Agents.** Academic Press, Inc.

Bertell, R. (1986). **No Immediate Danger? Prognosis for a Radioactive Earth.** Canadian Scholars Press.

Bertell. R. (2002). **Planet Earth: The Latest Weapon of War.** Quartet Books (UK).

Billac, P. (1999). **The Silent Killer: Indoor Air Pollution.** Swan Publishing.

Bosch, R. Van Den. (1989). **The Pesticide Conspiracy.** University of California Press.

Brimblecombe, P. (2011). **The Big Smoke: A History of Air Pollution in London since Medieval Times.** Routledge.

Bryson, C. (2006). **The Fluoride Deception.** Seven Stories Press.

Bullard, R. D., and M. Waters (2005). **The Quest for Environmental Justice: Human Rights and the Politics of Pollution.** Sierra Club Books.

Burdon, R (1999). **Genes and the Environment.** CRC Press.

Burdon, R. (2003). **The Suffering Gene: Environmental Threats to Our Health.** Zed Books.

Burlakova, E. B., and V. I. Naidich (2005). **The Effects of Low Dose Radiation: New Aspects of Radiobiological Research Prompted by the Chernobyl Nuclear Disaster**. VSP Books.

Butterworth, D. S. (2008). **The Radium Watch Dial Painters**. Lost Horse Press.

Button, G. (2010). **Disaster Culture: Knowledge and Uncertainty in the Wake of Human and Environmental Catastrophe** Left Coast Press.

Caldicott, F., et al. (1998). **Mental Disorders and Genetics: The Ethical Context**. Nuffield Council on Bioethics, London.

Carlo, G. and M. Schram (2002). **Cell Phones: Invisible Hazards in the Wireless Age: An Insider's Alarming Discoveries about Cancer and Genetic Damage**. Basic Books.

Carson, R. (2002). **Silent Spring**. [Anv edition.] Mariner Books. [Afterword: E. O. Wilson.]

Caufield, C. (1990). **Multiple Exposures: Chronicles of the Radiation Age**. University of Chicago Press.

Center for Health, Environment & Justice. (1991). **The American People's Dioxin Report**.

Chang, M. I. (2011). **Pollution in China**. Nova Science Pub, Inc.

Childs, B. (1999). **Genetic Medicine: A Logic of Disease**. The Johns Hopkins University Press.

Clark, C. (1997). **Radium Girls: Women and Industrial Health Reform 1910 – 1935**. The University of North Carolina Press.

Clegg, B. (2010). **Armageddon Science: The Science of Mass Destruction**. St. Martin's Press.

Colborn, T., et al. (1997). **Our Stolen Future: Are We Threatening Our Fertility, Intelligence, and Survival? A Scientific Detective Story**. Plume.

Cole, M. D. (2002). **Three Mile Island: Nuclear Disaster**. Enslow Publishers, Inc.

Collins, F. S. (2010). **The Language of Life: DNA and the Revolution in Personalized Medicine**. Harper.

Cook-Deegan, R. (1994). **The Gene Wars: Science, Politics and the Human Genome**. W.W. Norton.

Corburn, J. (2005). **Street Science: Community Knowledge and Environmental Health Justice**. The MIT Press.

Cracraft, J., et al. (1999). **The Living Planet in Crisis**. Columbia University Press.

Davis, D. H. (2007). **Ignoring the Apocalypse: Why Planning to Prevent Environmental Catastrophe Goes Astray**. Praeger.

Davis, D. L. (2003). **When Smoke Ran Like Water: Tales of the Environmental Deception and the Battle Against Pollution**. Basic Books.

Davis, D. S. (2009). **Genetic Dilemmas: Reproductive Technology, Parental Choices, and Children's Future** (Second Edition). Oxford University Press, USA.

Dhingra, G. (1987). **Pesticide mutagenesis: A tabulated review**. Environmental Mutagen Society of India.

Didenko, V. V. (2010). **DNA Damage Detection In Situ, Ex Vivo, and In Vivo: Methods and Protocols**. Humana Press.

Dimaio, M. S., et al. (2010). **Prenatal Diagnosis: Cases and Clinical Challenges**. Wiley-Blackwell.

Ehrlich, P. R., and A. H. Ehrlich (2009). **The Dominant Animal: Human Evolution and the Environment**. Island Press.

Epstein, C. J., et al. (2008). **Inborn Errors of Development**. Oxford University Press, USA.

Evans, H. J., and D. C. Lloyd (1979). **Mutagen-induced Chromosome Damage in Man**. Yale University Press.

Evans, M. A. (2010). **Wounded Earth** [Kindle Edition]. Joyeuse Press.

Evans, M. D., and M. S. Cooke (2010). **Oxidative Damage to Nucleic Acids**. Springer.

Fitzgerald, R. (2007). **The Hundred-Year Lie: How to Protect Yourself from the Chemicals That Are Destroying Your Health**. Plume.

Fradkin, P. L. (2004). **Fallout: An American Nuclear Tragedy.** Johnson Books.

Freeman, L. J. (1982). **Nuclear Witnesses. Insiders Speak Out.** W. W. Norton and Company, Inc.

Frickel, S. (2004). **Chemical Consequences: Environmental Mutagens, Scientist Activism, and the Rise of Genetic Toxicology.** Rutgers University Press.

Garrod, A. (1909). **Inborn errors of metabolism.** Oxford University Press.

Geacintov, N. E., and S. Broyde (2010) **The Chemical Biology of DNA Damage.** Wiley-VCH.

Goldman, M. I. (1975). **The Spoils of Progress: Environmental Pollution in the Soviet Union.** The MIT Press.

Goldstein, S., and C. R. Reynolds (2010). **Handbook of Neurodevelopmental and Genetic Disorders in Children** (Second Edition). The Guilford Press.

Gould, J. M., and B. A. Goldman. (1993). **Deadly Deceit: Low-level Radiation, High-level Cover-up.** Four Walls Eight Windows.

Greene, G. J. (2001). **The Woman Who Knew Too Much: Alice Stewart and the Secrets of Radiation.** University of Michigan Press.

Gregory, D. W. (2003). **Radium Girls.** Dramatic Pub Co.

Haffhold, S. E. (2011). **Encyclopedia of Water Pollution**. Nova Science Pub, Inc.

Hamer, D. and P. Copeland (1998). **Living with our genes**. Doubleday.

Hansen, J. (2010). **Storms of My Grandchildren: The Truth About the Coming Climate Catastrophe and Our Last Chance to Save Humanity**. Bloomsbury USA.

Harte, J., et al. (1991). **Toxics A to Z: A Guide to Everyday Pollution Hazards**. University of California Press.

Hawkins, M. (1997). **Social Darwinism in Euopean and American Thought**. Cambridge University Press.

Hayatsu, H. (1990). **Mutagens in Food: Detection and Prevention**. CRC Press.

Hill, M. K. (2010). **Understanding Environmental Pollution** (Third Edition). Cambridge University Press.

Hollaender, A., ed. (1976). **Chemical Mutagens** (Four volumes.)

Horne, R. (2010). **A Is for Armageddon: A Catalogue of Disasters That May Culminate in the End of the World as We Know It**. Harper Paperbacks.

Huddart, D., and T. Stott (2010). **Earth Environments: Past, Present and Future**. Wiley.

Ibrahim, A. M. (2010). **Soil Pollution: Origin, Monitoring & Remediation**. Springer.

Icon Group International (2010). **Genetics and Disease: Webster's Quotations, Facts and Phrases**. ICON Group International, Inc.

Ishihara, T., and M. S. Sasaki (1983). **Radiation-induced Chromosome damage in Man**. A. R. Liss.

Jacobs, C., and W. Kelly (2008). **Smogtown: The Lung-Burning History of Pollution in Los Angeles**. Overlook Hardcover.

Jamieson, D. (2008). **Ethics and the Environment: An Introduction**. Cambridge University Press.

Johannessen, O. M., et al. (2010). **Radioactivity and Pollution in Nordic Seas and Arctic: Observations, Modelling and Simulations**. Springer.

Kahn, M. E. (2006). **Green Cities: Urban Growth and the Environment**. Brookings Institution Press.

Kevles, D. (1985). **In the Name of Eugenics**. Harvard University Press.

Khoury, M., et al. (2010). **Human Genome Epidemiology**. Oxford University Press, USA.

Kimura, H., and A. Suzuki (2008). **New Research on DNA Damage**. Nova Science Publishers.

Knasmuller, S., et al. (2009). **Chemoprevention of Cancer and DNA Damage by Dietary Factors**. Wiley-VCH.

Knudson, A. G. Jr. (1900). **Genetics And Disease**. McGraw-Hill Book Company.

Lambin, E., and M. B. DeBevoice (2007). **The Middle Path: Avoiding Environmental Catastrophe**. University of Chicago Press.

Levine, M. J. (2007). **Pesticides: A Toxic Time Bomb in Our Midst**. Praeger.

Lupski, J. R., and P. T. Stankiewicz (2006). **Genomic Disorders: The Genomic Basis of Disease**. Humana Press.

Lynn, R. (1996). **Dysgenics: Genetic Deteriorations in Modern Populations**. Praeger Publishers.

Mahajan, B. S., and M. S. Rajadhyaksha (1999). **New Biology and Genetic Diseases**. Oxford University Press, USA.

Markowitz, G., and D. Rosner (2003). **Deceit and Denial: The Deadly Politics of Industrial Pollution**. University of California Press.

Markus, H. (2003). **Stroke Genetics**. Oxford University Press, USA.

Martin, P. (1997). **The Sickening Mind: Brain, Behaviour, Immunity, and Disease**. Harper Collins, London.

Mazur, A. (1998). **A Hazardous Inquiry: The Rashomon Effect at Love Canal**. Harvard University Press.

Mazzocco, M. M. M., and J. L. Ross (2007). **Neurogenetic Developmental Disorders: Variation of Manifestation in Childhood**. The MIT Press.

McKeown, W. (2003). **Idaho Falls: The Untold Story of America's First Nuclear Accident**. Ecw Press.

Medvedev, Z. A. (1980). **Nuclear Disaster in the Urals**. W. W. Norton and Company, Inc.

Medvedev, Z. A. (1992). **The Legacy of Chernobyl**. W. W. Norton & Company

Milunsky, A. (ed.), and J. Milunsky (ed.) (2010). **Genetic Disorders and the Fetus: Diagnosis, Prevention and Treatment**. (Sixth Edition). Wiley-Blackwell.

Mittica, P. (2007). **Chernobyl: The Hidden Legacy**. Trolley Press

Miura, S., and S. Nakano (2008). **Progress in DNA Damage Research**. Nova Biomedical Book.

Morris, R. D. (2008). **The Blue Death: The Intriguing Past and Present Danger of the Water You Drink**. Harper Paperbacks.

Mosley, S. (2008). **The Chimney of the World: A History of Smoke Pollution in Victorian and Edwardian Manchester**. Routledge.

Moynihan, R., and A. Cassels. (2006). **Selling Sickness: How the World's Biggest Pharmaceutical Companies Are Turning Us All Into Patients**. Nation Books.

Mullner, R. (1999). **Deadly Glow: The Radium Dial Worker Tragedy**. American Public Health Association.

Murgatroyd, C. (2010). **The Power of the Gene: The Origin and Impact of Genetic Disorders**. Nova Science Publishers, Inc.

Nadakavukaren, A. (2005). **Our Global Environment: A Health Perspective**. Waveland Press, Inc.

Nass, R., and Y. Frank (2010). **Cognitive and Behavioral Abnormalities of Pedriatic Diseases**. Oxford University Press.

Nesse, R. M., and G. C. Williams.(1996). **Why We Get Sick: The New Science of Darwinian Medicine**. Vintage.

Oeijord, N. K. (2000). **Human Instincts Explained**. Vantage Press.

Oeijord, N. K. (2002). **Genetic Catastrophe! Sneaking Doomsday?** iUniverse, Inc.

Oeijord, N. K. (2003). **A Dictionary of Genetic Damage**. iUniverse, Inc.

Oeijord, N. K. (2005). **Derailed Evolution**. iUniverse, Inc.

Pariza, M. W., et al. (1990). **Mutagens and Carcinogens in Diet**. Wiley-Liss.

Payne, R. (2010). **How Much is Enough?: Buddhism, Consumerism, and the Human Environment**. Wisdom Publications.

Pennington, B. F. (2008). **Diagnosing Learning Disorders** (Second Edition). The Guilford Press

Pritchard, D. J., and B. R. Korf (2007) **Medical Genetics at a Glance** (Second Edition). Wiley-Blackwell.

Fupecki, S. R. (ed.) (2006). **Genetic Screening. New Research**. Nova Science Publishers.

Rabinoff, M. (2010). **Ending the Tobacco Holocaust**. Elitebooks

Rachel Carson Council, Inc. (1992). **Basic Guide to Pesticides: Their Characteristics and Hazards**. CRC Press.

Rapp, D. J. (2003). **Our Toxic World: A Wake Up Call**. Environmental Research Foundation.

Reddy, P. R. (2006). **Chromosome Damage by Environmental Agents**. Hesperides Press.

Richards, B. J. (2006). **Fight for Your Health: Exposing the FDA's Betrayal of America**. Truth in Wellness.

Rosenberg, R. N., et al. (2007). **The Molecular and Genetic Basis of Neurologic and Psychiatric Disease**. Lippincott Williams & Wilkins.

Ross, B., and S. Amter (2010). **The Polluters: The Making of Our Chemically Altered Environment**. Oxford University Press, USA.

Sage, C. (2011). **Environment and Food**. Routledge.

Sanchez, M. L. (2008). **Causes and Effects of Heavy Metal Pollution**. Nova Science Publishers.

Sanders, H. J. (1969). **Chemical mutagens; The road to genetic disaster?** American Chemical Society.

Sarkar, B. (1999). **Metals and Genetics**. Springer

Schwela, D., and O. Zali (1998). **Urban Traffic Pollution**. Spon Press.

Shabecoff, A. (2008). **Poisoned Profits: The Toxic Assault on Our Children**. Random House.

Shaw, M. W. (1970). **Human Chromosome Damage by Chemical Agents**. Annual Review of Medicine. Vol. 21: 409 – 432

Sheffield, M. E. (2011). **Encyclopedia of Air Pollution**. Nova Science Pub, Inc.

Simmon, V. F. (1977). **Evaluation of selected pesticides as chemical mutagens 'in vitro' and 'in vivo' studies**. N. T. I. S.

Smil, V. (2008). **Global Catastrophes and Trends: The Next Fifty Years**. The MIT Press.

Smith, R., and B. Lourie. (2010). **Slow Death by Rubber Duck: The Secret Danger of Everyday Things**. Counterpoint.

Sokhi, R. S., and M. Molina (2011). **World Atlas of Atmospheric Pollution**. Anthem Press.

Sonntag, C. von (2010) **Free-Radical-Induced DNA Damage and Its Repair: A Chemical Perspective**. Springer.

Spellman, F. R. (2009). **The Science of Environmental Pollution** (Second Edition). CRC Press.

Steen, R. Grant (2010). **Human Intelligence and Medical Illness: Assessing the Flynn Effect**. Springer.

Stonehouse, B. (2009). **Arctic Air Pollution**. Cambridge University Press.

Takada, J. (2005). **Nuclear Hazards in the World**. Springer.

Thomas, A. E. (2010). **DNA Damage Repair: Repair Mechanisms and Aging**. Nova Science Publishers

Thomas, S. (1986). **Genetic Risk**. Pelican, Lonon.

Tucker, T. (2010). **Atomic America: How a Deadly Explosion and a Feared Admiral Changed the Course of Nuclear History**. Bison Books.

Valverde, C. (2010). **Genetic Screening of Newborns: An Ethical Inquiry**. Nova Science Pub, Inc.

Verchick, R. R. M. (2010). **Facing Catastrophe: Environmental Action for a Post-Katrina World**. Harvard University Press.

Wexler, A. (1995). **Mapping Fate**. University of California Press.

Wilson, E. O. (1999). **Consilience: The Unity of Knowledge**. Vintage

Wilson, E. O. (2003). **The Future of Life**.

Wilson, E. O. (2007). **The Creation: An Appeal to Save Life on Earth**. W. W. Norton & Company.

Withgott, J. H., and S. R. Brennan (2010). **Environment: The Science Behind the Stories** (4th Edition). Addison Wesley.

Wynbrandt, J., and M. D. Ludman (2008). **The Encyclopedia of Genetic Disorders and Birth Defects** (Third Edition). Facts on File.

Yablokov, A. V., et al. (Four editors) (2010). **Chernobyl: Consequences of the Catastrophe for People and the Environment (Annals of the New York Academy of Sciences)**. Wiley-Blackwell.

Yaroshinskaya, A. et al. (1995). **Chernobyl. The Forbidden Truth**. Bison Books.

Zehnder, A. J. B. (1995). **Soil and Groundwater Pollution: Fundamentals, Risk Assessment and Legislation**. Springer.

Index

Note: The INDEX refers to Chapter / Section and it only refers to the Chapter / Section where the entry is first mentioned.